Newts in your pond and garden

by

James Grundy

Sincere thanks to all my family, friends and colleagues who offered me encouragement and support. Special thanks go to my mother, father and wife Carole. Without them this book would never have been started, let alone finished.

Male palmate newt

Published by James Grundy 3 Harper Close Macclesfield Cheshire SK11 7QG

Tel: 01625 869921
Email: grun@macc111.freeserve.co.uk

ISBN
0-9553775-0-1
978-0-9553775-0-1

©Copyright in the text and photographs James Grundy 2007

All rights reserved.

This book is protected by copyright. No part of this publication may be reproduced, stored in a retrieval system or transmitted in any form or by any means, electronic, mechanical, photocopying, recording or otherwise without prior written permission of the copyright holder.

A catalogue record for this book is available from the British Library

Edited by Gillian Pierce and Carole Grundy

Book cover design by Jason Lawton

Printed in the UK by Buxton Press Ltd

This book is a 2007 first edition

Contents

Introduction
page 4

The pond
page 6-14

The garden
page 15-20

The great crested newt
page 21-36

The smooth newt
page 37-51

The palmate newt
page 52-65

Establishing newts in your pond
page 66-69

Monitoring and recording
page 70-76

Observing and handling newts
page 77-82

What do you know about newts?
page 83-88

Useful Contacts
page 89 -92

Introduction

My reason for writing this book is a desire to record and share what I have learned about our native newts through direct, hands on experiences over the last forty years. Its purpose is to give others a practical source of help and advice so that they can encourage these fascinating creatures to establish themselves in their own ponds and gardens.

Like me, you probably remember the times you spent as a child at your local pond, collecting frogspawn or catching newts and tadpoles and bringing them home in a motley array of containers to varying degrees of parental approval. Then, assuming they did not escape overnight, spending many happy hours engrossed in caring for them and marveling at the development of frogspawn and newt eggs; first into tadpoles and then tiny baby frogs or newts no bigger than a fingernail. For some this led to an ongoing and general interest in natural history, and for me a voluntary role working for the conservation of our native amphibians, a real job as a night driver for Royal Mail and from August 2006 paid employment with an ecological consultancy working on amphibian conservation projects.

It is widely acknowledged that all three species of newt native to the United Kingdom are under threat from our continued development of open spaces and modern agricultural methods. Sadly many of the ponds that we frequented in our youth have been lost to housing developments, out of town shopping centers or new roads. The frogs, toads and newts that used to breed in them either killed or forced to search, often unsuccessfully, for new breeding sites. The result has been a big reduction in the populations of all our native newts, with one species; the great crested or, as I used to know it, the warty newt, now considered to be rare and endangered. The outlook for all our newt species was looking grim until, unexpectedly, about twenty years ago, a virtual revolution in how we garden began.

This revolution, which continues today, was gardening for wildlife and as the eighties became the nineties, ponds, water features and ways of encouraging wildlife into our gardens became increasingly popular and fashionable, fuelled by gardening books, magazines and mainstream TV programmes like Gardeners World and more recently Spring Watch.

As the number of wildlife friendly gardens has grown, amphibians from the surviving wild populations have started to colonise them, with the common frog and smooth newt proving to be especially adept at establishing themselves in our ponds and gardens. Despite this positive turn of events, all newts, especially the great crested, can still find it very difficult, sometimes impossible, to reach otherwise perfectly suitable sites or, having found their way into a garden, can find it hard to survive and establish a new colony simply because the well-meaning owner lacks a basic understanding of their behavior and habitat requirements. This book aims to address this problem and encourage newt conservation by showing all gardeners and pond owners, how to;

- Create a newt friendly pond and garden,

- Attract newts into your pond and garden,

- Identify and sex all three native species,

- Recognise their eggs, tadpoles and young,

- Observe their behaviour and life cycle.

Even a small, well managed garden is quite capable of supporting an astonishing variety of wildlife. With just a little thought and effort most gardens can be designed and maintained to attract our native amphibians. Many suburban and town garden ponds already support large, breeding populations of frogs and smooth or palmate newts. Some gardens have even been colonised by great crested newts much to the surprise and delight (though not always!) of the owners.

1: A wildlife pond, butyl lined 2m x 4m x 90cm deep.
Amazingly, this small pond in a town center garden supports a breeding colony of thirty great crested newts, over two hundred smooth newts, about one hundred common frogs and a host of other pond creatures.

If your garden or pond is already home to great crested newts or is colonised by them in the future you must make sure that you fully understand the legislation relating to them. In simple terms it is currently an offence under both UK and European legislation to intentionally or recklessly kill, injure, take, or disturb great crested newts at all stages of their life cycle. It is also an offence to intentionally or recklessly damage, destroy or obstruct access to places of shelter or protection used by them, this includes their breeding ponds as well as terrestrial habitats. The only exceptions are actions carried out under the appropriate licence granted by the relevant licensing agency. In certain circumstances the presence of great crested newts within 500m of a proposed development can result in severe planning constraints and considerable expense to progress building projects, (this can include house extensions, new driveways and garages!). Breaking the laws relating to great crested newts could result in a fine of up to five thousand pounds and or a period of up to six months imprisonment for each offence committed. As wildlife legislation is often amended it is important that you regularly review the current situation by contacting the relevant licensing agency (see page 89) for up to date advice about great crested newt legislation. Remember ignorance of the law is no defence!

Chapter one

The Pond

Newts in the garden pond

To understand the importance of the garden pond to our native newts you need to know that all adult newts, in common with frogs and toads, have to return to water in early spring to mate and lay their eggs. The majority of adult newts instinctively return to their birth pond and will spend about four months of every year, from March through to June, in the pond.

During this water-based (aquatic) phase the adult newts' skin is regularly shed and undergoes a major physical change becoming semi-permeable to oxygen. This allows them to absorb some oxygen directly from the water and in simple term's means, that in certain conditions, they can survive under water by breathing through their skin. This ability is very important, particularly if the breeding pond freezes over in spring, as it enables early returning adult newts to survive being trapped under ice at the bottom of the pond for prolonged periods, without drowning. However, adult newts cannot remain submerged indefinitely and will eventually drown if they are continually prevented from surfacing, or the dissolved oxygen content of the water falls to dangerously low levels.

1a: A female smooth newt in the pond, preparing to rise to the surface to take air.
In warm weather an adult newt must surface about every ten minutes to breathe.

As the water temperature starts to rise with the advance of spring, and the danger of ice recedes, the adults become increasingly active and an increased metabolic rate triggered by longer days and warmer weather means they must regularly come to the surface to breathe. During warm weather you will see adult newts repeatedly swimming up through the water column and taking a gulp of air at the surface. Often surfacing is accompanied with a distinctive popping sound, as the newt takes a breath.

While in the pond adult newts and their larvae, more commonly known as newt tadpoles, feed regularly. They take a wide variety of live prey items including frog tadpoles, insect larvae and invertebrates (water fleas and midge larvae are particular favourites). All prey is caught using the mouth, then immediately swallowed whole. Among all newt species, especially in ponds with large populations, cannibalism is normal and adult newts will indiscriminately predate their own and each others eggs and tadpoles.

2: Great crested newt tadpoles, approx ten weeks old, hunting for prey.
The tadpoles of all our native newt species take about sixteen weeks to complete their development from egg to young newt. All newt tadpoles are cannibalistic; the larger tadpoles will often attack and eat smaller siblings.

Newt eggs and the resulting tadpoles can only survive in water and usually take about sixteen weeks to complete their metamorphosis into young newts. All newt tadpoles initially breathe by taking oxygen directly from the water, using a set of external gills sited on either side of the head.

By about two weeks old the tadpoles' internal lungs are developing and they start rising to the surface to take gulps of air. However the external, feathery looking gills are retained and remain functional until the tadpoles leave the pond as young newts in late summer early autumn, at which time the gills are absorbed into the body.

To understand why newts and other amphibians seem to prefer some gardens and ponds over others you need to know what they require, in terms of aquatic and terrestrial habitat, why these requirements are important and how you can employ simple garden management and pond maintenance regimes to encourage newts and other wildlife to colonise your garden naturally.

Pond location

A pond intended for newts should be located in a sunny part of the garden, preferably where it will receive full sun from February through to October, for as much of the day as possible. The pond should be well clear of shady overhanging trees and shrubs. If this is unavoidable, it may be necessary to cover it with netting every autumn to prevent fallen leaves fouling the water. Direct sunlight is important to all our native newt species for four reasons;

- The warmth provided by direct sunlight triggers courtship, mating and egg laying in the adults,

- Water warmed by the sun speeds the rate of newt egg embryo development,

- Warm water encourages feeding activity in the resulting newt tadpoles,

- Direct sunlight powers the growth of microscopic plants and algae in the pond providing food for a wide range of insect larvae and invertebrates, which are in turn predated by the adult newts and their tadpoles.

3: The location of the pond in the garden is very important.
This pond receives full sunlight for most of the day from early spring into autumn.

Correct edging is an important part of any pond. Getting it right can be a matter of life and death for young newts, who may drown if they are unable to get out of the pond when they have absorbed their external gills. It is essential that all young newts have a way out of the water directly into the cover of an 'emergence zone' that extends around at least a third of the pond. This zone provides them with shelter from predators during their first critical 24/48 hours on land, giving them time to adjust to the transition from a water living tadpole, to land based (terrestrial phase) young newt.

Emergence zone

The emergence zone is a strip of permanently damp, un-maintained land around your pond. Stretching from the pond edge out into the surrounding garden for about a metre, this strip should be allowed to grow wild and over time a wide range of native plant species will naturally colonise it.

To help speed the colonisation process you can, if you wish, introduce a mixture of your personal wetland plant favourites. The encroachment of grasses from the zone into the pond will provide its inhabitants with easy access in and out of the water and help soften the pond's outline hiding any straight lines and corners.

4: The emergence zone extends right around this pond and softens its straight edges. The zone extends from the pond edge out into the surrounding lawn and contains a mixture of various grasses, rush, marsh marigold, buttercups, clover, willow herb, nettles and docks.

The more plant varieties in the emergence zone, the more attractive and useful it becomes to insects and other wildlife in your garden. It provides young newts emerging from the pond in late summer with a good source of food and shelter. All newts, adults and young, on leaving the pond, will spend several days resting and feeding in this zone before dispersing into the surrounding gardens.

Water level management

All garden ponds naturally go through cycles of losing water by evaporation and gaining it through rainfall. To maintain the water at a particular level it is a good idea to keep a couple of large buckets permanently filled with tap water and use them to top up as required. Tap water left outside for at least twenty-four hours will 'naturalise' and any chemicals present, like chlorine, will disperse into the atmosphere. Overfilling of the pond sometimes caused by heavy rain will cause the ponds inhabitants no major problems, while the emergence zone and surrounding garden should soak up any overflow.

In crisis situations, for example when the water level becomes very low, it is generally acceptable to top up the pond using a hose and water direct from the tap. However you need to be aware that in some regions the chemicals and trace elements found in freshly drawn tap water might have an adverse effect on some of the plants and animals living in your pond.

Pondweed management

The choice of pondweeds and water plants suitable for garden ponds is extensive and ranges from the British natives through to a wide range of cultivated and hybrid varieties. Fortunately, newts are quite versatile and they will usually make use of whatever plants are available in your pond.

As the pond will form a key part of your garden, it makes sense for you to use water plants that fit in with your personal preferences and situation. If you make a 'mistake' in your choice of plants or want a change, you always have the option to thin, remove or replace. Removal of any pond weed should be undertaken after the newt breeding season, to prevent newt eggs being destroyed, and care taken not to injure or kill any newt tadpoles sheltering within the weed. Proven newt friendly pondweeds and water plants include; hornwort, milfoil, starwort, canadian pondweed, curled pondweed, water forget-me-not, water mint, brooklime, water plantain and all the flote grasses.

Pondweeds and water plants are very important in the pond because they oxygenate and remove nutrients from the water, while providing the ponds inhabitants with food, cover and a place for newts and other pond creatures to lay their eggs. Most pondweeds and water plants can either be purchased from garden centres, specialised suppliers or begged from friends and family with established ponds.

Duckweed, if introduced into your pond, can quickly become a major problem because of its vigorous growth; especially in the smaller garden ponds, without ongoing management it will rapidly cover the waters surface and damage the pond's ecosystem and food webs by blocking out vital sunlight. Duckweed can be kept under control in a small pond by regularly skimming it off with a fine meshed hand net and leaving it on the bank to dry out.

Fast growing blanket weed can also cause problems, especially in new ponds. Blanket weed can be controlled by slowly rotating a bamboo cane in the water, so that the thin strands of this filamentous weed wrap themselves round the cane, much like candyfloss on a stick, enabling you to remove it from the pond with the minimum of disturbance and effort. Again care should be taken not to inadvertently remove newt eggs or tadpoles hidden within the weed.

A general tidy up of the pond and emergence zone should be undertaken every November. This involves removing any dead plant material and all the pondweed from about a third of the pond's total area. This will help to create open spaces in the pond and let in that all important sunlight. To give any creatures caught up in the removed pondweed a chance to get back into the pond, lay it onto a chicken wire frame and suspend it over the water for forty eight hours before composting.

Fish in the pond

The advice from most wildlife organisations is that fish and newts simply don't mix and that under no circumstances should fish ever be introduced into ponds used by or intended for newts. While I agree with this advice, it does need to be mentioned that a proportion of garden ponds containing fish also have small colonies of smooth or palmate newts present.

In general, a garden pond newt colony with under ten breeding age adults should be considered a borderline population that has a high risk of becoming extinct. Very often small newt populations in garden ponds containing fish are not self-sustaining and the colony is maintained by the arrival of adult newts from bigger breeding colonies in neighbouring ponds. Occasionally, to confuse the issue even further, the specific conditions in some fish ponds will allow just enough newt eggs and tadpoles to survive to maintain a relatively small but viable newt colony.

Despite the existence of these mixed fish/newt ponds and the apparent contradiction they appear to present, it is important to understand that all the commonly available species of garden pond fish will feed voraciously on both the eggs and tadpoles of all our native newts. Even small numbers of fish in a pond are quite capable of predating most, if not all, the newt tadpoles from a season's spawning. The fish will also take invertebrates and insect larvae, leaving adult newts and any surviving tadpoles with very little or nothing to feed on.

Many factors come into play when fish and newts share a pond, these include:

- The size and number of fish in the pond,
- If the fish are regularly fed or left to fend for themselves,
- The species and number of adult newts using the pond before fish were introduced,
- Levels of fish and newt predation,
- The proximity of other newt colonies,
- The availability of green corridors linking the site to neighbouring newt colonies,
- Number of newt tadpoles successfully reaching maturity,
- The ponds size, weed cover and range of habitats.

Verified reports of great crested newts sharing ponds with fish are unusual and the loss of several large great crested newt colonies has been directly linked to the illegal introduction of fish, accidental or otherwise, into their breeding ponds.

Sadly, it is often well intentioned but ill informed members of the public who, in what they see as an act of kindness, move fish from one pond to another or release ornamental fish into the wild, with disastrous consequences for the resident native amphibian populations. Once introduced, fish are notoriously difficult to remove from a pond and often the best option, to resolve the problem, is to create a new, purpose built, fish free breeding pond within 250m of the original breeding site.

Pond life colonisation

From the moment you start to fill a new pond with water, various species of animals, insects and invertebrates will start to arrive. They employ a wide variety of methods to reach new sites, often blown in on the wind or hidden among the water plants you beg or buy. Many of the tiny, wriggling, unusual looking creatures you see in the water are the larval stages of airborne insects, these include; gnats, midges, mosquitoes, stonefly, damsel and dragonflies. Frogs are especially opportunistic and will quickly find their way unaided into most garden ponds. These arrivals are perfectly normal and form a key part of a ponds natural development and continue for as long as the pond holds water. A good way of 'kick starting' or speeding up this colonisation process is to collect and add a bucket or two of water, pondweed and mud from an established, fish free, garden pond in your area.

5: Water from an established pond.
Pond water is a living soup of invertebrates and insect larvae including; flatworms, snails, water fleas, pea mussels, stonefly, fresh water shrimp, midge larvae and water louse.

It has been estimated that a single pond can provide a home to over two thousand different species. These range from microscopic plants and animals through to insects, invertebrates, fish, amphibians and small mammals like the water vole. To help with species identification it is essential to have a good reference book detailing the creatures you are likely to find. I use a pocket sized copy of the Observers book of Pond Life by John Clegg; its clear illustrations, descriptions and photos usually puts me on the right track when I find something new in the pond. Before you and your family can start to learn first hand about this amazing variety of pond life you will need to design and construct your new garden pond or consider the management of an existing pond to encourage them into your garden.

Wildlife pond design and construction

A wildlife pond in your garden is one of the best ways to introduce the next generation to the natural world and stimulate their interest in conservation issues. It is an unparalleled educational resource and all the pond's inhabitants, (especially the newts!), will provide you and your family with years of enjoyment and could set you or your children on the road to a lifetime's involvement with nature. For most garden ponds, particularly the small to medium sized, I recommend using the black butyl liners now widely available from most Garden Centres and DIY stores. These robust, flexible liners are supplied with easy to follow instructions on pond construction, should last for at least fifteen years, and will enable most people to shape and profile a pond of their own design.

6: Duckweed on the surface of a garden pond.
Avoid introducing duckweed into your pond, it will rapidly cover the surface, blocking out vital sunlight.

To attract smooth and palmate newts, your pond needs to be at least two metres by one and a half metres, with a depth of at least sixty centimetres over two thirds of its area. To attract great crested newts, your pond should be at least four metres by two metres, with a minimum depth of ninety centimetres over two thirds of its area. All our native newts naturally prefer still or very slow moving water. Therefore powerful water pumps, fountains, waterfalls and filtration systems should be avoided or, if used, only switched on occasionally for short periods.

Your local library, Wildlife Trust and the Internet will have books, leaflets and sites full of pond design and construction ideas. You should have a look at as many options as possible and use a design that fits in best with your garden and personal preferences. As a starting point, and to help you avoid some of the pitfalls with a new pond, I have included a few tips based on my own experience and mistakes.

Tips on pond construction

- Decide what you are going to do with the soil excavated to create your pond, options include a skip to remove it completely or using it as part of a landscaping project elsewhere in the garden.

- Decide on the pond size and shape, and mark it out on the surface of the soil using string or sand. Ensure your local supplier has the correct size liner available.

- Cut and stack some grass turfs from the site of the pond or elsewhere in the garden; cut enough to edge the whole pond and create a natural pond bottom.

- Excavate the pond to the required depth and preferred shape. Profiling of the sides (shaping) should include at least one shallow water "shelf".

- Line the bottom and sides of the excavation with a two centimetre thick layer of sand, newspapers or old carpet to create a protective layer between the liner and soil.

- Carefully lay the Butyl liner on top of this protective layer; easier if done by two people. Take care not to puncture or tear the liner.

- Before filling the pond with water, ensure that all the liner up to the level of the pond edge will be covered with water; use a spirit level and a straight edge laid across the length and then width of the empty pond to check the horizontal levels.

- Using a hose and tap water, fill the pond to almost overflowing. Your aim is to have all the liner completely immersed, up to the level of the surrounding garden.

- When satisfied with the water level, carefully put some of the previously cut turfs into the pond, grass side up, to create a natural pond bottom.

- The remaining turfs can now be used to edge the pond. Place them directly onto the excess Butyl liner around the edges of the pond. They will hide and hold it in place. The various species of grass and plants growing in the turf will form part of the emergence zone plant mix.

- Folds and creases in the liner are unavoidable and provide useful cover for the ponds inhabitants. The creases are quickly hidden by plant growth as the pond establishes.

- After filling with water, and before introducing plants, the pond should be left for a couple of weeks to settle and naturalise. At the end of the second week the pond and the emergence zone can be planted up with your selection of water plants, pondweed and edging plants.

- From time to time the water might turn a bright green, especially in new ponds, this is normal and caused by algal growths or 'blooms' fuelled by sunlight, warm weather and dissolved nutrients in the water. The pond will eventually clear as the algal blooms die off or are consumed by colonising insect larvae and invertebrates.

- When finished you can sit back, pour a drink and enjoy the fruits of your labour. A new pond, depending on its size, should only require minimal maintenance in the first two or three years of its life. Unfortunately you cannot rest for long, because as adults, all our native newts only spend about a third of the year in the pond. The rest of the time they are in their terrestrial phase living in and around the garden, so its time to get busy again.

Chapter two

The garden

Newts in the garden

To understand the importance of the garden to our native newts, you really only need to know two things; first, all young newts spend their first three years living almost exclusively away from the pond. Second, all adult newts spend up to eight months of every year out of the pond living in the garden and its surrounds.

7: Newt friendly habitats in a town garden.
Garden habitats include old walls, neglected corners, flower borders, the compost heap and lawns.

After leaving the pond, the skin of the smooth and palmate newt becomes dry and leathery in both texture and appearance while the skin of the great crested newt develops a rough warty look with distinctive ebony like sheen. In their terrestrial phase the ability of an adult newt to absorb oxygen directly through its skin is lost and will not be regained until it returns to the pond to breed again the following spring. These skin changes are an important survival strategy for all adult newts as they help to protect them from drying out; this allows them to range over a wide area in search of shelter, prey, new breeding sites and explains why they occasionally turn up in gardens without ponds.

With just a little thought and effort, even small gardens can be managed to provide the necessary range of habitats to support a newt colony. When assessing the suitability of your garden, don't forget to include neighbouring gardens in your deliberations, as 'your newts' will be using them as well. Most suburban and town gardens with their flower borders, neglected corners, old stone walls, rockeries, compost heaps and wide range of garden plants make a perfect home for wildlife and newts.

Plants in the garden

When deciding on plants for your garden, bear in mind the need to attract a steady supply of live insect prey for the newts in their terrestrial phases. You should be looking to plant various species of nectar producing plants that flower in succession from early spring into late autumn, with some plants specifically chosen to provide leafy shade and ground cover. Most untreated lawns are useful for providing newts with a reliable source of earthworms, as are; compost heaps, log piles (especially if partially buried in the ground), neglected corners, perennial borders and annual flowerbeds. The wide range of plants available from garden centres, specialist suppliers and plant catalogues, should allow you to achieve a result that fits in with your personal preferences, the practicalities of your garden and the need to attract insects.

8: Neighbouring gardens can provide a variety of newt habitats.
Newts from your pond will feed and shelter in neighbouring gardens; they also use them as green corridors to disperse over a wide area.

A good tip is to visit your local garden centre once a month throughout the spring and summer to see what's in flower in your area and, most importantly, attracting the insects. If you buy one or two plants each time you should be able to guarantee a succession of flowers for you to enjoy and a good supply of insect food for the newts. All gardens, even those designed to attract wildlife, have to be maintained on a regular basis. With newts in residence a little extra care is needed when working in the garden.

Garden maintenance

Garden maintenance poses a number of dangers to newts and other amphibians in their terrestrial phases; however the main threats come from, lawnmowers, grass strimmers and projects that involve landscaping works. To help minimise the risks, lawns should be kept short and regularly mown, ideally in the late afternoon after a hot sunny day, when most amphibians will be under cover. If the lawn becomes overgrown it will need to be 'walked' before cutting and any amphibians found moved into safe, suitable cover away from the area to be mown.

Landscaping projects should, if possible, be put off until the winter with care taken not to disturb winter shelters used by the newts (see page 20). If a project cannot be delayed it is worthwhile regularly searching the site for newts before starting work each day. It will help to keep the newts away if the site is kept tidy and clear of debris that they might use for shelter.

Without being alarmist, it is important to consider that if great crested newts are living in your garden, picking them up and moving them, and seemingly routine garden and pond maintenance tasks, could be viewed as deliberate or reckless capture and disturbance. I would therefore recommend that anyone with great crested newts living in their garden contact the relevant licensing agencies for further advice.

9: A newt friendly perennial flower border.
Old bricks and half-buried logs hidden in the flower border will provide extra shelter for your newts.

A mixed flower border, with nasturtium, phlox, golden rod, lavender, various herbs, perennial wallflowers, sweet peas and a mixture of low growing flowering shrubs will provide your newts with a good hunting ground and plenty of undisturbed summer shelter. With some forward planning the border should require little maintenance and attract a steady supply of insects into the garden.

Weed and pest control

The best and simplest advice is to use no weed killers or chemical pest controls of any description. Weeding should be done the old fashioned way, with a hoe or by hand, and the insects you previously battled and regarded as pests such as aphids and black fly should be viewed as a useful source of food for your newts. Once newts have become established in your garden these pests will be kept under control by newt predation and other natural predators, such as ladybirds, hoverflies, lacewings and predatory wasps encouraged into the garden by your 'green' gardening methods. In some circumstances, for instance, ant swarms, wasp nests and plagues of slugs or snails, you may feel the use of some chemical control is the only option. In general, newt populations as a whole will tolerate the very limited and localised use of things such as ant poison, wasp killer, slug pellets and spot applied glyphosate based weed killers providing great care is taken not to contaminate the pond.

10: Recently emerged from the pond, a young smooth and great crested newt.
Young newts can often be found hunting for insects on and around the compost heap.

Compost heaps

All gardens should have a compost heap to help manage garden and organic household waste. Aside from the obvious environmental benefits, a compost heap provides newts with a useful source of insects, invertebrates, earthworms and a place to shelter. The best designs are the wooden slatted or open fronted types; these allow for easy newt access in and out of the heap. Care should always be taken when adding or removing compost material, to avoid injuring any newts that might be sheltering or feeding in the heap. In their terrestrial phase, all newts, adults and young, must have access to two sorts of cover. The first, known as a 'summer shelter' is any sort of cover that gives them a safe, undisturbed, cool, daytime retreat out of the sun from April through to October.

Summer shelter

A summer shelter for newts can be provided by a range of sites, including; thickly planted flower borders, long grass, a pile of old rubble, compost heaps, old stone walls, nooks and crannies in old brick work, undisturbed log piles and spaces under paving slabs and large stones. All newts will instinctively seek out and may gather together in these areas, as it begins to come light in the morning, following a night's hunting in the garden. Individual newts will often return to the same favoured summer shelter time after time.

11: August, can you see all five smooth newts sheltering from the sun?
A few old bricks or logs left undisturbed in the flower border will make a good summer shelter for your newts; unfortunately slugs, like the one above, will also use them and may reward your kindness by eating your plants!

Summer shelters are very important to all newts, as they help protect them from drying out in hot weather and keep them safely hidden from a range of predators such as cats, hedgehogs and birds. Occasionally, usually after rain, newts will emerge during the day in search of prey especially if the weather has been dry for a prolonged period or the night time hunting has been poor.

The second type of cover, required by newts from November to March, is called a 'winter shelter'. As the name suggests this is any sort of cover that gives them a safe, undisturbed, frost-free refuge where they can spend the winter.

Winter shelter

A good winter shelter will provide newts with a secure place to spend the winter months, protected from predators and the hard frosts that can kill them. They instinctively seek out this type of shelter in late October early November, favoured locations include; down among the foundations of old brickwork, deep inside old stone walls, compost heaps, piles of rubble, small mammal burrows, or partially buried log piles. If necessary, you can easily build a perfectly acceptable winter shelter in your garden by digging out a trench two metres long by sixty centimetres wide and forty-five centimetres deep. This trench should be loosely backfilled with material from the compost heap mixed with old bricks, stone, rotted timber, logs or clean builder's rubble and then capped off with old paving stones. The aim is to create a warren of linked tunnels and cavities below ground. The un-mortared gaps and cracks between the old paving slabs should give the newts' easy access to this subterranean shelter.

A purpose built winter shelter should be sited in a dry, sheltered part of the garden and must be left undisturbed throughout the winter. Take great care not to construct the shelter in a frost pocket or area likely to become waterlogged after heavy rain. Repeated flooding and freezing of the underground cavities during the winter will kill any newts sheltering there.

12: Old stone walls provide excellent winter shelter for all newts.

The gaps and cracks in old stone or brick walls provide newts with easy access to a safe, frost-free winter shelter deep inside the wall and down among its foundations. Once they have taken cover, for the winter, all newts stop feeding and enter a dormant or torpid state, similar to hibernation in mammals. This dormant state normally begins with the first hard frosts in November and lasts until the weather starts to warm the following spring, when the adult newts emerge from their winter shelters and begin to make their way back to the breeding ponds.

All our native newts have very similar behaviour and life cycles and the differences between them, are often quite subtle. The following chapters look at each of our three native newt species in more detail, starting with the great crested newt.

Chapter three

The Great Crested Newt
(Triturus Cristatus)

General information

The great crested newt is the biggest and rarest of our three native species and it is strictly protected by law. Before you can catch, handle or disturb the adults, eggs or young or disturb their places of terrestrial shelter and breeding ponds you must have a licence, depending on location, either from; Natural England (NE), Countryside Council for Wales (CCW) or Scottish National Heritage (SNH).

Great crested newts seem to prefer larger, fish-free ponds and in the past they proved to be very successful at colonising abandoned open cast mineral workings, including old sand, gravel and clay quarries. The largest known great crested newt colony in the world, (approximately 30, 000 animals) is to found living in the old clay workings at Orton Pit in Peterborough. The species can be found in locations right across the United Kingdom, though its main strongholds are in East Anglia, North East Wales and the North West of England.

The species takes its name from the males' distinctive jagged crest which develops only in the breeding season during the aquatic phase, and their much greater size in comparison to the other two species. The skin of both males and females is dark brown, shading through to black, with hard-to-see dark spots and a rough, warty appearance. Adult and young newts have white stippling around the head and along the flanks, and orange banding or 'socks' on the front and rear toes. The underbelly is always a bright yellow/orange with irregular black blotching. The purpose of this striking belly colour is to warn predators of the newts' poisonous skin. If taken by a predator or roughly handled specialist glands just below the skins surface release foul tasting, noxious toxins to stop and deter attacks.

The adults become sexually mature at three to four years old when they are about ten centimetres in length. As with all newts, they continue to grow throughout their life, consequently the largest individuals will normally be the oldest. Female great crested newts can reach eighteen centimetres in length and are always larger than same age males, sometimes by as much as a third. Adult newts can live for up to fifteen years in suitable garden habitats, reportedly longer in captivity. Where great crested newts occur, it is usual to find them sharing the site with one, or occasionally both, of the other two native newt species.

The adults feed regularly while in the pond, taking live prey made up of a wide range of pond dwelling invertebrates and insect larvae. In common with all newts, they will also opportunistically predate their own and each others eggs and young. Adult great crested newts will enthusiastically take freshly chopped earthworms dropped into the pond for them.

It is widely accepted among conservationists that the great crested newt has been the least successful of our native newts at colonising urban ponds and gardens. Though in truth, the subject is very under researched and little hard data is available. Evidence of great crested newt presence in garden ponds is unreliable and based mainly on reports sent in by members of the public to the various conservation organisations. However, it is possible that many people have decided, for various reasons, against reporting the presence of great crested newts in their pond. As a consequence the actual number of garden ponds currently supporting breeding populations of great crested newts is unknown.

Great crested newt identification

Males in the pond

Male great crested newts usually start to return to the breeding ponds in late February. This return of sexually mature males continues throughout March and sometimes extends into April, depending on the temperature and weather conditions. On first entering water, the male's distinctive jagged crest is barely visible and it can take several weeks to fully develop. Eventually reaching as much as three centimetres in height, the crest runs from head to tail down the middle of the males back with a break or notch where the tail joins the body. The whole purpose of the breeding crest is to impress the females and signal to them his readiness to breed and suitability as a mate. The crest is supported and held upright by the water, collapsing to lay flat against his back if he is removed from the pond.

As the males come into breeding condition their tails enlarge and flare, the white stripe on both sides of the tail lightens becoming more pronounced and the cloaca, sited at the base of the tail between the hind legs, blackens and swells.

13: April, a male great crested newt in the pond, approx 12cm in length.
Identified as a male by his impressive breeding crest and enlarged flared tail with white central stripe.

By June the males have normally started to come out of breeding condition. Their breeding crest is gradually absorbed into the body and it shrinks to just a few millimetres in height. The swollen cloaca shrinks and the tail loses its flared, fleshy appearance and thins. When their crest has been absorbed, the males leave the pond, and resume their terrestrial life.

Females in the pond

Female great crested newts start to return to the breeding ponds in late February early March, as with the males, the return of the females can extend into April depending on the temperature and weather conditions. The females enter the water with bellies already heavily swollen with the season's eggs. It is believed that the majority of females in a breeding colony return to the pond a few days after the males, though conclusive evidence for this is currently unavailable.

Female great crested newts are easily identified in the pond by their large size, egg swollen bellies, lack of a crest and their tail which is always much thinner than the males and lacks the white stripe. Most females have a narrow yellow or orange band running along the bottom edge of their tail. This band remains visible throughout the year and can be a useful aid in the sexing of adult newts. Like the males the females have extensive dark body spots, which can be difficult to see, especially in the darker skinned individuals.

The female newts advertise their willingness to mate by releasing a cocktail of scents and pheromones from their cloaca into the water. These scents and the sight of a heavy female moving through the water will stimulate the males into pursuit and triggers their elaborate courtship display.

14: April, a female great crested newt in the pond, approx 17cm in length.
Identified as a female by her large size, egg swollen belly, lack of a crest, and the thin tail with a band of yellow/orange along its bottom edge.

By July, with mating and egg laying finished, the females become much sleeker and they begin to leave the breeding pond and disperse into the surrounding gardens. As the adult newts move onto land and enter their terrestrial phase the males and females start to look very similar to each other, and differentiating between the sexes becomes more difficult.

Adult great crested newts in their terrestrial phase

Great crested newts are mainly nocturnal in their terrestrial behaviours. They spend the days hidden away in summer shelters, only emerging after dark to hunt for earthworms and other live prey. On land, the adults' warty skin takes on an ebony like sheen and glands on its surface exude a pungent, persistent scent. As they move around the garden at night these glands create scent trails that they can use to find their way back to preferred daytime shelters.

15: August, a male great crested newt during his terrestrial phase, approx 12cm in length.
The warty nature of the great crested newts' skin is clearly visible in the terrestrial phase. The adult newt in the above photograph was identified as a male by the remnant of its breeding crest and the white stripe on the tail. Great crested newts have very good night vision and an excellent sense of smell.

The best way to sex a terrestrial phase adult great crested newt is to check its tail. You are looking for the male's distinctive white tail stripe, which remains visible all year round. The females' tail lacks this white stripe and, though less conspicuous, it should have a narrow, unbroken band of orange/yellow along its bottom edge.

Another good way of sexing terrestrial phase adult great crested newts is to look closely at their backs. Adult males always retain a just-visible remnant of their breeding crest and a careful examination will reveal the crest running from head to tail lay flat against the back.

An individual newt's success, or otherwise, in hunting and finding summer shelter during the terrestrial phase from July to October, impacts directly on its breeding condition the following spring. For example, a well fed female will return to the pond with a bigger egg load than an undernourished female, while a healthy, well fed male will develop a bigger more impressive breeding crest when he returns to the pond, increasing his chances of attracting a female. Adult newts that have poor feeding opportunities during the terrestrial phase will be so stressed that they may not survive the winter. At best they will return to the pond in such a poor condition that their chances of breeding successfully are greatly reduced.

16: September, a female great crested newt during the terrestrial phase, approx 17cm in length. Identified as a female by her large size, no remnant crest and lack of white stripes on the tail.

From late October the adults instinctively seek out winter shelters. Locations vary, but deep inside old stone walls or down among the foundations of old brick walls are favourite places for great crested newts to spend the winter. Often several individuals will find and share the same site. As the days shorten and the temperature starts to fall the adult newts stop feeding and enter a torpid state. Care should be taken not to disturb them during this period as exposure to a hard frost can kill them. Several days of unseasonable mild winter weather will sometimes rouse them from their torpor and stimulate them into emerging to hunt for food, and can occasionally trigger a risky early return to the breeding pond.

Great crested newt breeding behaviour

Sexually mature adults of both sexes begin the journey back to the pond from their winter shelters during February and March. They move mostly at night and particularly after rain when the temperature is above 10°C. Cold snaps or a strong chilling wind brings the return to a temporary stop with the newts taking shelter until it warms up again. Breeding begins as the water temperature starts to rise in March and continues through to June. Courtship and mating is a complicated process that can only take place under water. The following description will give you an idea of the process and help you to identify breeding behaviour if you are lucky enough to see it for real in your pond.

17: A male great crested newt in pursuit of a female during the first stage of courtship. The male shows the yellow, dark blotched underbelly common to both sexes.

The cloaca, referred to in the following description, is a multifunction body opening present in both male and female newts of all species. It is used to produce scents and pheromones during courtship, for egg laying, spermatophore transfer and to pass bodily wastes. The cloaca is situated at the base of the tail between the newts' hind legs.

- Courtship and breeding starts with the male great crested newt pursuing the female and scenting her cloaca with his nostrils. He does this to confirm that she is of the same species, female and receptive to his advances.

- After confirming the female's receptivity, the male takes up a position in front of her and displays by arching and bending his body to show his crest and body markings to best effect. At the same time he gently fans scents released from his cloaca towards her with his tail, while slowly backing away from her.

18: May, a male great crested newt displaying.
A unique, female eye view, of the males courtship display.

- When suitably impressed and stimulated the female starts to follow him around the pond while he continues to posture, display and fan.

- After several minutes of this following behaviour, the male turns to face away from her, raises his tail and releases from his cloaca, a white, sticky, elongated packet of sperm about 2mm in length. This packet of sperm is called a spermatophore.

- He then turns back to face the female and resumes the display leading her forward. As she moves over the spermatophore it attaches to her cloaca and is taken up inside her body where the sperm is released and fertilizes a portion of her eggs.

Both sexes will mate with multiple partners over the course of a season, however not all matings are successful as the spermatophore sometimes fails to attach correctly to the females cloaca and is lost. Occasionally more than one male will try to court the same female at the same time, in the resulting confusion the courtship display usually breaks down and the spermatophore transfer does not take place. Following her first successful mating the female great crested newt begins to lay her eggs.

Great crested newt eggs

Egg laying via the females cloaca starts within a few days of her first successful mating and peaks during April and May. Female great crested newts attach each sticky-coated egg individually onto the leaf of a suitable water plant and then use their powerful hind feet to fold it around the egg; its sticky coating secures the folded leaf in place hiding the egg from hungry eyes.

Most great crested newt eggs are laid at night under cover of darkness in submerged aquatic vegetation just below the pond's surface, with the female taking several minutes to carefully position and wrap each egg. A female newt will often examine the leaf of a plant by 'nosing' it before deciding to lay onto it.

The bigger the female the more eggs she will produce, the larger females can lay up to three hundred eggs over the spawning season and will take about twelve weeks to lay and wrap them all. However because of a genetic abnormality within the species half of all the eggs laid by the females never hatch. Great crested newt eggs are oval, four to five millimetres in diameter, with a clear jelly coating and a large, distinctive, creamy white nucleus.

19: April, several newly laid great crested newt eggs, approx 4mm in diameter.
Great crested newt eggs have a distinctive white nucleus inside a clear jelly shell. Half of all the great crested newt eggs laid every season are non-viable and there is no possibility of them developing into newt tadpoles.

By folding the leaf around the egg the female great crested newt hopes to hide and protect it from the unwelcome attention of predators including other newts. It is not unusual for the same plant to be used repeatedly by one or more females, resulting in water plants and pondweeds with numerous, multi-folded leaves.

Due to the prolonged breeding season it is quite normal to find great crested newt eggs at various stages of development in the pond, often on the same plant, at the same time. In ponds used for breeding by more than one newt species it is sometimes possible to find their eggs on the same water plants and pondweeds.

Great crested newt egg development

The rate of great crested newt egg development is governed by temperature, with warm weather speeding it up and cold slowing it down. Great crested newt eggs laid in March when the water in the pond is cold can take up to four weeks to develop into newt tadpoles while those laid later in the season, when the water is relatively warmer, only take about two weeks to become free swimming tadpoles.

The crested egg nucleus starts its life as a creamy white oval inside a clear jelly shell. As the cells in the nucleus divide it gradually elongates forming the head, body and tail of an embryonic newt tadpole. Continuing to develop and grow, but constrained by the eggs jelly shell, the developing tadpole gradually curls itself round inside the egg. Eventually it forms itself into a circle, with the embryonic tadpoles head touching the tip of its tail.

20: May, four developing crested newt eggs actual size approx 4mm.
The flote grass blades have been carefully teased open to reveal the developing eggs. The distinctive white coloration of the great crested newt egg nucleus is maintained as it develops.

In the days just prior to the embryonic crested newt tadpole emerging from the egg, its skin colour changes becoming a light yellowish green, and dark camouflage stripes form along the length of its head, body and tail. The developing tadpole now begins to vigorously twist and turn inside its jelly shell and in direct sunlight this rapid movement appears to make the egg flicker. Eventually the egg's jelly shell splits open and the tadpole swims off to take it's chances in the pond. Great crested newt tadpoles normally take about sixteen weeks to complete their metamorphosis into young newts.

Great crested newt tadpole development

Week one

Immediately after breaking free of the egg, great crested newt tadpoles seek shelter among the pondweed. There they rest for several hours using up the last reserves from the nucleus before starting to hunt for microscopic invertebrates. During these first weeks of life, they are dependant on their coloration and hiding among the pondweed to protect them from predators.

21: May, a great crested newt tadpole, approx 5mm in length.
Recently emerged from the egg and resting on a submerged leaf of water mint.

The newly hatched great crested newt tadpoles are very small, only four to five millimetres in length. They start to feed within hours of emergence and given warm weather and a good supply of live prey they will grow very quickly. They emerge from the egg with large, fully formed eyes, essential for hunting, but no limbs.

They breathe using a set of feathery external gills situated on either side of the head. Using their tail they swim in quick short bursts, with frequent stops among the pondweed to rest. At this stage of their development the tadpoles resemble fish fry and are very fragile, quickly dying if they suffer injury or are removed from the water.

During these first few weeks of life the vast majority of great crested newt tadpoles become a meal for a wide range of the pond's predators including water beetles, backswimmers and dragonfly/damselfly nymphs. This heavy predation continues until the tadpoles are about twelve weeks old when poison glands under their skin start to develop making them unpalatable and deterring most predators.

Week four

At four weeks old the great crested newt tadpoles are about two centimetres in length and the front legs are just visible. They occasionally rise to the surface to take a gulp of air to aid with buoyancy and to supplement the oxygen absorbed from the water through their external gills.

In a distinct change of behaviour they now start to emerge from the pondweed and take to floating in areas of clear open water hunting for prey, especially water fleas (daphnia). This behaviour, not shared with other newt tadpoles, makes them particularly vulnerable to predators and is believed to be the main reason why this species is so susceptible to fish predation.

22: June, a great crested newt tadpole four weeks old and approx 2cm in length.
The front legs have now started to grow though the individual toes have not yet formed. Its tail has fleshed out and a dark speckling is starting to appear across its head, body and tail.

Injuries at this age can often prove fatal to young great crested newt tadpoles as their immune systems are not yet fully formed, even minor damage to gills, newly formed limbs or the tail, can trigger bacterial and fungal infections quickly leading to their death.

Week eight

At eight weeks old the great crested newt tadpoles are about four centimetres in length. They have a heavy, thickset appearance, a large fleshy tail and fully formed front and rear legs complete with individual toes. Their external gills are clearly visible as large feathery appendages on either side of the head. Though still using the gills to absorb oxygen from the water, their internal lungs are well developed and they will rise to the surface every few minutes to take a gulp of air. If a great crested newt tadpole is prevented from surfacing for a prolonged period it will eventually drown.

23: July, a great crested newt tadpole eight weeks old and approx 4cm in length.
The tadpole's skin is dark mottled with a greenish brown hue; at this stage the underbelly is unmarked and silvery white. All newt tadpoles have superb underwater vision and an excellent sense of taste.

They have now become fearsome pond predators in their own right and, in the absence of fish; they have little to fear from the ponds other inhabitants. Their immune systems are fully formed and they are no longer susceptible to infections. Great crested newt tadpoles are voracious predators and they will attack almost anything smaller than themselves, that they perceive as prey.

Great crested newt tadpoles employ a slow stealthy approach when hunting, this cautious approach is followed by a lightning-quick lunge, to catch prey items with their mouth. As great crested newt tadpoles lack any teeth all captured prey is swallowed whole and live, anything unpalatable or to big to swallow is quickly spat out.

Week twelve

At twelve weeks old the great crested newt tadpoles are about five centimetres in length and still feeding voraciously. Poison glands are starting to form under their skin, as these glands develop the skin thickens, becomes darker, and gradually takes on the distinctive warty appearance of the species. The external gills are still present and functioning but have started to shrink as they are gradually absorbed into the body. The gills will continue to get smaller until they finally disappear completely. Once absorbed the gills cannot be re-grown and are lost forever.

24: August, a great crested newt tadpole twelve weeks old and approx 5cm in length.
Missing toes on the front foot are characteristic of a sibling attack. The developing poison glands in the skin make the tadpole unpalatable and deter most predators. The tail has lost its fleshy appearance and started to thin. The whole appearance of the tadpole has become more newt like.

The tadpoles now hunt either on the pond bottom or among the pondweed, as they have become too heavy to float easily in open water. They begin to spend increasingly longer periods resting and basking in the shallows around the edge of the pond or among the pondweed close to the surface.

By about twelve weeks old great crested newt tadpoles have started to develop the yellow, black blotched underbelly distinctive of the species.

These belly markings, once fully formed at about three years old, are believed to be permanent and unique to the individual newt, much like human fingerprints. This has proved useful to conservationists and ecologists at sites where continuous monitoring of great crested newts involving their repeated capture, is taking place.

At some important breeding sites hundreds of digital photographs of adult great crested newt bellies have been taken to build up comparative databases. These photographed newts, when recaptured, are revealing a wealth of data on their movements, distribution and life span.

25: August, a great crested newt tadpole, twelve weeks old and approx 5cm in length.
Rising to the surface to take a gulp of air. The young newt's yellow belly and black blotching is a warning to potential predators.

The development of the great crested newt tadpole's yellow, black blotched underbelly coincides with the development of the poison glands in the skin. This bold coloration, common to many species of animals and insects, acts as a warning and deterrent to potential predators; "don't eat me, I am poisonous". Predators quickly learn that at this stage of their development great crested newt tadpoles taste horrible and should be left alone!

Week fourteen

At fourteen weeks old, most great crested newt tadpoles are on the verge of leaving the pond. They have stopped feeding and their external gills have almost been completely absorbed. Most of their time is now spent resting at the surface of the pond or in the weedy shallows.

26: August, a great crested newt tadpole approx fourteen weeks old and 6cm in length.
The external gills, having shrunk down to small stubs, are just visible on either side of the tadpoles head.

Every so often the tadpoles will pull themselves free of the water and take a few hesitant steps into the emergence zone before returning to the pond. This behaviour continues for a couple of days until the remnants of the external gills are completely absorbed. Once the gills are lost they cannot be re-grown and the tadpoles leave the pond and begin their new life, on land, as young great crested newts.

After leaving the pond, usually at night or after rain, the young newts will normally spend a few days living in the emergence zone before moving out into the garden in search of food and shelter. This is a dangerous time for young great crested newts, and in these first few weeks, many are lost to lawnmowers, drainage grids, unwary gardeners and predators yet to learn about their poison glands.

Week sixteen

By sixteen weeks old most great crested newt tadpoles have left the pond and become young newts. They will not return to the pond until they are old enough to breed in three to four year's time. Over the course of these years, a few young newts will travel several kilometres and, with luck, may establish new breeding colonies in neighbouring ponds. Most young great crested newts are much less adventurous, and they will often live out their entire lives within a two hundred-metre radius of their birth pond.

27: September, a young great crested newt sixteen weeks old, approx 6cm in length.
The external gills have been absorbed into its body and the poison glands under the skin are fully developed. A young great crested newt's skin colouration, ebony sheen, and warty look exactly matches that of the adult newt.

Through August and September the survivors of the year's young continue to leave the pond and disperse into the garden and its surrounds. They are about six centimetres in length and perfect miniature versions of the adults. Young great crested newts spend their days under cover and the nights, before winter begins, hunting for small insects and earthworms.

During late October early November, as the days shorten and the temperature starts to fall, they instinctively seek out a frost-free winter shelter; adult and young great crested newts will often find and share the same winter shelters. With the first frosts the young newts stop feeding and enter a torpid state. They will emerge as the weather starts to warm the following spring, and resume their hunt for prey around the garden.

Chapter Four

The Smooth Newt
(Lissotriton Vulgaris formally Triturus Vulgaris)

General information

The smooth newt is the most widespread of our three native newt species. It can be found in locations right across the United Kingdom and is believed to be the only newt naturally native to Ireland. Sometimes called the common newt, it has proved to be very successful at colonising suburban gardens and ponds. Given the right conditions, populations of over a hundred adult smooth newts, are not unusual even in small ponds. This species is known to co-exist with fish in some garden ponds, but only in much reduced numbers.

Smooth newts reach sexual maturity at approximately three years old when they are about six centimetres in length. They continue to grow throughout their lives and can live for up to twelve years, with the largest individuals normally the oldest. Females, who are always bigger than same age males, can reach ten centimetres in length.

Within the species there is a wide range of colouration, markings and size differences between individual newts of the same and opposite sex. They do not possess poison glands and lack the ability to produce distasteful toxins, consequently they are frequently taken by a wide range of predators, (both in and out of the pond), including diving beetle larvae, dragonfly nymphs, fish, cats, dogs, herons, foxes, badgers and hedgehogs.

The species takes its name from the smooth, slippery nature of the adult newts' skin during their aquatic phase. On first returning to the pond, the adult newt's skin is dry and leathery in appearance but within a few hours this layer is shed to reveal the new 'smooth' moist, skin beneath. The adults usually eat the shed skin, but, on occasion, you may find it floating in the water or among the pondweed as a filmy, ghostly white newt shape. In their aquatic phase the adult newts' skin is semi-permeable and allows them to absorb some oxygen directly from the water; this is especially useful for early returning adults, particularly if the pond freezes over for short periods.

Smooth newts feed regularly while in the pond taking mainly live prey made up of pond dwelling invertebrates, frog tadpoles, insect larvae and, opportunistically, their own eggs and young. In ponds shared with breeding colonies of frogs, the adult smooth newts become especially skilled at biting through the frogspawns jelly covering to get at the developing tadpoles. This can result in the total loss of an entire season's frogspawn, especially in ponds with a large newt population.

Due to similarities in size and appearance smooth and palmate newts are frequently mistaken for each other, because of this a careful visual examination of any newts found in your pond or garden, to ensure their correct identification, is always recommended. If you already have newts living in your garden they are probably smooth newts. That said, it is not unusual for more than one species of newt to share the same breeding pond and terrestrial habitat.

Smooth newt identification

Males in the pond

Male smooth newts start to return to the breeding ponds in March. This return is spread over four to six weeks, with the last males arriving back at the pond in April. On first entering water the males have no breeding crest and it can take several weeks for it to fully develop.

The breeding crest eventually reaches around one centimetre in height and runs in an unbroken, wavy line from head to tail down the middle of the male's back. In the pond the crest remains upright, out of water; it collapses to lay flat against the male's back.

28: May, a male smooth newt in the pond, approx 7cm in length.
This male clearly shows the breeding crest, dark 'leopard' body spots, feathered or splayed rear toes and the flared tail with a broken band of colour along its bottom edge.

As the male comes into breeding condition his tail flares and a broken band of colour develops along its bottom edge. The toes on the hind feet become feathered or splayed and the cloaca, sited at the base of the tail between the hind legs, becomes black and swollen. A key feature for identifying male smooth newts, both in and out of the pond, are the dark 'leopard' spots covering the upper body.

Females in the pond

The female smooth newts start to return to the breeding ponds in March, as with the males their return is spread over about six weeks. The females arrive back at the pond with their bellies already heavily swollen with the season's eggs.

29: April, a female smooth newt in the pond, approx 9cm in length.
Note the female's uniform brown skin colouration, heavy egg swollen appearance, lack of 'leopard' body spots, and the un-feathered rear toes.

Female smooth newts can usually be distinguished from the males in the pond quite easily. Characteristics to look for include; their egg swollen bellies, a dull uniform skin colour; which ranges from light through to dark brown, the lack of a breeding crest, the absence of any 'leopard' body spots, a much thinner tail, and by their rear toes, which are not feathered or splayed.

Both in and out of the pond the female smooth newt's tail remains thin and usually has a narrow band of orange running along its bottom edge. The female's underbelly is normally off white with a pale orange central stripe and is freckled with small, dark, irregular spots.

29a: April, a female smooth newt in the pond, approx 9cm in length.
As she rises to the surface to take a gulp of air, the lighter freckled underbelly with its pale orange, central stripe and her cloaca at the base of the tail are clearly visible.

As the breeding season comes to an end in July, the adults undergo a number of physical changes in preparation for leaving the pond. The male smooth newts absorb their breeding crests, their hind toes lose the splayed appearance and their enlarged, flared tail thins. The female smooth newts become much sleeker as they finish laying their eggs.

After leaving the pond, at the end of the breeding season, the skin of both sexes becomes dry and leathery and its ability to absorb oxygen directly from water is lost. On land the males and females closely resemble each other. Differentiating between male and female smooth newts in their terrestrial phases can sometimes be quite difficult.

Adult smooth newts in their terrestrial phase

In their terrestrial phase, from July to October, adult smooth newts spend the daylight hours under cover in a variety of summer shelters, emerging after dark or rain to hunt for earthworms and insects around the garden. As with the other newt species, success in hunting and finding shelter during this period impacts directly on their breeding success the following spring.

30: September, an adult male smooth newt during his terrestrial phase, approx 8cm in length. Identified as a male by the 'leopard' body spots.

On land the males' key identifying features are; 'leopard' body spots, which remain visible throughout the year, a remnant of the breeding crest running the length of the back and a cream, off white underbelly, which retains its dark spotting, and red, central stripe.

After leaving the pond, the females' key identifying features are; a uniform dull brown colouration, the absence of dark 'leopard' body spots, the narrow band of orange running along the bottom edge of the tail and a cream, off white underbelly with dark freckling and pale orange central stripe.

The dry, leathery nature of the adult smooth newt's skin and its uniform colour in the terrestrial phase sometimes results in individual newts being mistaken for lizards. The giveaway for correct identification is the smooth newts' habit of freezing and playing dead, even allowing you to pick them up. Common lizards, which rarely turn up in gardens, are usually far less cooperative and dart away at high speed when disturbed, giving you little chance of even seeing them, let alone catching one!

Adult and young smooth newts, of both sexes, look very similar to each other in their terrestrial phase. Differentiating between the sexes can be quite difficult when you find smooth newts in the garden away from the pond.

31: September, adult and young smooth newt approx 10cm and 4cm.
The adult was identified as an old female by her size, leathery, dull brown skin colour, the lack of 'leopard' body spots and no remnant of a breeding crest. The smaller newt is probably a two year old immature female, although sexing immature young newts correctly is always difficult.

Visual observations of the smooth newts living in your pond and garden can be supplemented with some careful handling and reference to the photographs in this book. With practice and experience you will quickly become proficient at identifying the species and sex of any newts that you find in and around your garden.

The two male smooth newts in the photograph below, both found in the same garden, exhibit the wide range of individual colouration present within the species. They highlight the difficulty of correctly identifying and sexing adult smooth newts in their terrestrial phase. They were sexed as males after a careful examination revealed that they both had 'leopard' body spots and the just visible remnant of a breeding crest.

32: September, two adult male smooth newts during their terrestrial phase, approx 7.5cm in length. They highlight the potential difficulties in sexing adult smooth newts found in the garden and show why they are sometimes mistaken for lizards.

From late October the adult smooth newts instinctively seek out frost-free winter shelters around the garden. Locations include cracks/gaps in old walls, log piles, compost heaps and purpose built shelters; often several individuals, (adults and young), will find and share the same site. With the onset of winter and the first frosts they stop feeding and enter a torpid state, similar to hibernation. Care should be taken not to disturb known winter shelters between November and February as exposure to a hard frost can kill them. The following spring, as the weather warms, the adult smooth newts will emerge and begin their journey back to the breeding ponds.

Extended periods of unseasonable warm winter weather, (seemingly more frequent these days), can sometimes rouse smooth newts from their torpid state, and as with the other newt species, trigger a risky, early return to the pond.

Smooth newt breeding behaviour

Over a six-week period, from March through to April, adult smooth newts of both sexes begin the journey back to the breeding ponds from their winter shelters. Most adult newts return to the pond on wet nights, when the temperature is above 10°C. Cold weather or chilling winds will bring this return to a temporary stop. During a cold snap, the adult newts will shelter in any available cover until it warms up again, if they are unable to find adequate cover they run the risk of being killed by a hard frost or foraging predators. Smooth newt courtship and mating begins as the water temperature starts to rise with the onset of spring, and continues through to July. As with all our native newt species, courtship and mating can only take place in water.

The stages of smooth newt courtship and mating;

Chase and Scent,
The male chases after the female and scents her cloaca to confirm that she is the right species, sex and receptive to his advances.

Display and Fan,
The male positions himself in front of the female and displays by arching and twisting to show his breeding crest and colours to best effect. At the same time he uses his tail to vigorously fan scents, from his swollen cloaca, towards her as he slowly backs away.

Following,
When suitably impressed, the female starts to follow the male around the pond, while he continues to display and fan.

The Spermatophore,
After several minutes of following, the male turns to face away from the female, raises his tail and releases a sticky packet of sperm called a spermatophore from his cloaca. He then turns back to face her and continues to display and fan leading her forward.

Fertilization.
As the female passes over the spermatophore it attaches to her cloaca and is taken up inside her body where the sperm is released and fertilizes a portion of her eggs.

The courtship display of male smooth newts is far more vigorous and frenetic than that performed by male great crested newts and often takes place during the day. Smooth newt tail fanning is probably best described, as tail lashing and frequently several males will simultaneously chase after the same female. In the resulting melee, the males will often display to each other until they realise their mistake. Both sexes will mate with multiple partners during the breeding season, though not all matings are successful.

As smooth newts are normally active in the day, if you stand quietly by the side of the pond, especially on a bright sunny day, you should be able to watch the male smooth newts chasing the females and observe their elaborate courtship behaviour. With a little luck and patience you might even get to see the complicated process of spermatophore transfer. Following their first successful mating the female smooth newts begin to lay their eggs.

Smooth newt eggs

The female smooth newts start to lay their eggs (via the cloaca,) within a few days of their first successful mating and egg laying reaches its peak during April and May. The females attach each sticky-coated egg individually onto the leaf of a suitable water plant and then use their hind feet to fold the leaf protectively around each one. Most eggs are laid during the day, in submerged vegetation, just below the surface of the pond. A female smooth newt can take up to ten minutes to carefully position, lay and wrap each egg. The larger females will lay up to two hundred eggs over a season and can spend up to twelve weeks laying them all in this time-consuming way. The eggs are oval, approximately three millimetres in diameter, with a sticky, clear jelly coating. Smooth newt eggs can normally be identified by their size and the nucleus, which is usually a grey/brown colour.

33: April, smooth newt eggs approx 3mm in diameter, laid onto the blades of flote grass.
Some of the grass blades have been carefully teased open to show the eggs and the coloration of the nucleus within the eggs clear jelly shell.

Folding the leaf around the egg is intended to hide and protect it from predators. Often the same plant will be used repeatedly by one or more females resulting in water plants and pondweed with numerous multi-folded leaves. It is normal to find smooth newt eggs at various stages of development in the pond, even on the same plant, at the same time (because of the prolonged breeding season). In ponds, occupied by more than one newt species, it is sometimes possible to find water plants, and even leaf folds, containing the eggs of two or even all three newt species.

On bright, warm, sunny afternoons in spring or early summer, it is often possible to watch the female smooth newts laying. Given a little luck, you will see them, just below the water's surface, among the pondweed, rolling, twisting and exposing their orange underbellies as they attach their eggs to the pondweed. After laying their eggs and wrapping them in a leaf fold, the females show no further interest in them. The eggs are left to their own devices, either to develop into tadpoles or become a meal for hungry predators.

Smooth newt egg development

The rate of smooth newt egg development is governed by temperature, with warm weather speeding it up and cold slowing it down. Eggs laid in March can take up to four weeks to develop into tadpoles while those laid later in the season, when the water is warmer, only take about two weeks.

The egg nucleus starts life as a small, brown oval inside a clear, jelly shell. As the cells in the nucleus divide it gradually elongates forming the head, body and tail of an embryonic smooth newt tadpole. Continuing to grow, but constrained by the jelly shell, the developing tadpole gradually curls itself round inside the egg, forming itself into a circle, with the tadpole's head touching the tip of its tail.

34: May, four developing smooth newt eggs, approx 3 mm in diameter.
Showing, from left to right, the development into an embryonic tadpole. The embryo in the third egg is turned to show its lighter underbelly and confusingly it looks like a developing great crested newt egg.

In the days just before a smooth newt tadpole emerges from the egg, dark horizontal camouflage stripes form along the length of its head, body and tail and it starts to vigorously twist and turn. Eventually, the egg's jelly shell splits open and the tadpole breaks free and swims off into the pond. Given warm weather and a good supply of live prey the newly emerged tadpoles will develop very rapidly. Smooth newt tadpoles normally take about sixteen weeks to complete their development (metamorphosis) into young newts.

Smooth newt tadpole development

Week one

Newly hatched smooth newt tadpoles are very small, only about three millimetres in length. They already have fully formed eyes, used for hunting prey, but lack legs. They breathe using a set of external feathery gills, situated on either side of the head. Young smooth newt tadpoles will quickly die, if removed from the water or injured.

34a: May, a newly emerged smooth newt tadpole, approx 3mm in length.
This smooth newt tadpole is waiting for unsuspecting microscopic invertebrates to swim past. It will take them with a quick snap of the mouth. By taking cover among the pondweed, in this case hornwort, it hopes to avoid becoming a meal itself.

Immediately after breaking free of the egg, smooth newt tadpoles seek shelter among the pondweed. There they rest for several hours, using up the last reserves from the nucleus, before starting to hunt for microscopic invertebrates. In these first weeks of life they rely on their colouration and hiding among the pondweed to protect them from predators, even so, less than one in a hundred smooth newt tadpoles that reach this stage of development will survive to become adults.

Week four

At four weeks old smooth newt tadpoles are about 1.5cm in length and they have developed a pair of front legs. Though the legs are not yet strong enough to support their weight out of water, they are used to help the tadpoles move through the pondweed in search of prey.

The young tadpoles are still highly vulnerable to predation and instinctively avoid open water. They tend to keep to the weedy areas of the pond in the hope of avoiding hungry beetles, fish, nymphs, adult newts and older siblings. This behaviour is believed to be the reason why this species of newt is able to coexist with fish in some ponds.

34b: June a smooth newt tadpole, four weeks old and approx 1.5cm in length
This smooth newt tadpole is preparing to dine on the unsuspecting Cyclops. The tadpole's front legs and external gills are visible; the back legs are just beginning to form and can be seen as the two stumps either side of the tail.

At this age the young smooth newt tadpoles' internal lungs are well developed and they frequently rise to the surface to take a gulp of air to supplement the oxygen absorbed from the water through their external gills. Given a good supply of food and warm weather the tadpoles will continue to develop rapidly. Within two weeks of growing their front legs, their back legs will have started to develop and become visible as two stumps (buds) where the tail joins the body.

Week eight

At eight weeks old smooth newt tadpoles are about two centimetres in length, and they are usually a uniform light brown colour. Their front and rear legs, complete with toes, are almost fully formed. They use their legs to clamber through the pondweed and walk across the pond bottom in search of prey, which can include their younger siblings. At this age smooth newt tadpoles still absorb oxygen directly from the water through their external gills, but must also regularly rise to the surface to take gulps of air, if prevented from surfacing they will eventually drown.

Smooth newt tadpoles spend most of their time among the pondweed hunting for small invertebrates. They are still very vulnerable to predation and instinctively avoid open water and rely on keeping hidden and their colouration to avoid detection by predators.

35: July, a young smooth newt tadpole about eight weeks old and approx 2cm in length.
Over wintering in the garden pond is a common occurrence among late hatching smooth newt tadpoles.

Late hatching tadpoles from smooth newt eggs laid at the end of the breeding season do not have sufficient time to complete their development into young newts before winter arrives. So, triggered by shorter days and falling temperatures, they suspend their development and remain in the pond throughout the winter. The following spring, as the days lengthen and the water temperature starts to rise, these over wintering tadpoles resume their development. They will leave the pond, as young smooth newts, in early June of the following year. This is a very useful survival strategy, also employed by the other newt species, that ensures no newt eggs or tadpoles are unnecessarily wasted. This adaptation gives all over wintering newt tadpoles (if they can avoid predators) a clear advantage over the first of next season's young.

Week twelve

At twelve to fourteen weeks old smooth newt tadpoles are about three centimetres in length. Their external gills are shrinking, as they are gradually absorbed into the body. Their skin begins to thicken and takes on a brown, almost leathery appearance. Both front and rear legs are now fully developed and capable of supporting a tadpole's weight out of water.

35a: August, a smooth newt tadpole about fourteen weeks old and approx 3cm in length.
This smooth newt tadpole is resting among the pondweed at the water's surface. The external gills are just small stubs on either side of the head. The dark line down the center of the tadpole's back is a common feature of all newt tadpoles and young newts (all species); it disappears soon after they leave the pond.

At this age smooth newt tadpoles start to congregate in the weedy shallows; at the surface and around the edge of the pond. They will occasionally leave the water completely and venture into the emergence zone, after resting for a few minutes they return to the pond. This behaviour continues for several days until the external gills have been completely absorbed.

After the gills are lost the tadpoles must have easy access to the emergence zone around the pond, otherwise there is a real risk of drowning. Skin coloration among individual smooth newt tadpoles and young terrestrial phase newts varies considerably, even among siblings, and ranges from dark through to light brown.
.

Week sixteen

At sixteen weeks old, and about three centimetres in length, most smooth newt tadpoles have left the pond and become young smooth newts. In common with young great crested newts they will not return to water until old enough to breed in about three year's time. During these land-based years a few individuals will travel several kilometres, if not lost to predators or accident, they may establish new breeding colonies in neighbouring ponds. The majority of young smooth newts spend their first three years living within a two hundred metre radius of the birth pond and will return there to breed.

36: September, a young Smooth newt sixteen weeks old and approx 3cm in length.
This young smooth newt's skin is light brown, (almost yellow), dry and leathery. The small, dark line on the neck marks the position of the absorbed external gills.

Through August and September the survivors of the season's spawning continue to leave the pond and disperse into the garden and its surrounds. They spend the days hidden away and the nights hunting for small insects on the compost heap, in the flower borders and across the lawn. In October/November, as the nights lengthen and the temperature starts to fall, young smooth newts join the adults in seeking out frost-free winter shelters around the garden. With the onset of winter and the first frosts they stop feeding and enter a torpid state. The following spring, as the weather warms, they will emerge along with young newts from previous years and resume hunting for prey in the garden.

Chapter Five

The Palmate Newt
(Lissotriton Helveticus formally Triturus helveticus)

General information

The palmate newt is the smallest of our three native newts, it can be found in locations right across the British Isles with the exception of Ireland. This species of newt is believed to prefer the slightly more acidic ponds found on heathland, uplands, peat bogs, and within woodland. The palmate newt also appears to have been quite successful at colonising garden sites, with some ponds and water features, known to support breeding populations of well over one hundred adults.

Palmate newts reach sexual maturity at three years old, when they are about five centimetres in length. Like all our native newts, they continue to grow throughout their life; consequently the larger individuals are normally the oldest. The females, who are always bigger than same age males, can reach up to eight centimetres in length. Palmate newts can live for about twelve years in garden habitats, slightly longer in captivity.

Within the species there are wide variations in size, skin colour and markings between individuals of the same, and opposite sex. These variations make a close examination to correctly determine the species and sex of individual newts essential. Palmate newts take their name from the male newt's distinctive, permanently webbed hind feet which turn black, and are especially conspicuous during the breeding season.

In the pond the skin of both male and female palmates becomes semi-permeable, enabling the adult newts to take dissolved oxygen directly from the water. This ability shared with other newt species allows adult newts returning to the pond in early spring to survive trapped under water should the pond ice over for short periods.

At the end of the breeding season, when the adult palmate newts leave the pond, their skin changes and its semi-permeable qualities are lost. On land, in their terrestrial phase, the adult's skin is dry and leathery in texture and appearance.

Adult palmate newts feed regularly while in the pond and take a wide variety of live prey including invertebrates, insect larvae and frog tadpoles. In their terrestrial phase they hunt in the garden for small insects, aphids and earthworms. Palmate newts do not posses poison glands and lack the ability to produce toxins to deter predatory attacks, consequently, fish, cats, wildfowl, foxes, badgers and herons can take a heavy toll on the adult population.

Palmate newts are sometimes found sharing the same breeding ponds and terrestrial habitats with one, or, occasionally, both of the other two native newt species. Like all newts they will opportunistically predate their own and each other's eggs, and young. The true extent of their success in colonising garden ponds is uncertain, mainly because the species is frequently misidentified and confused with the smooth newt.

Palmate newt identification

Males in the pond

Male palmate newts start to return to the breeding ponds in March, with the return of individual newts spread over a six-week period. The male palmates key identifying features in and out of the pond are their distinctive, black, permanently webbed rear feet. In sharp contrast, to the male newts of other species, the male palmates breeding crest, even when fully formed, is barely visible. The male's crest consists of a rigid, unbroken, raised ridge of flesh about three millimetres high running from head to tail down the middle of the back. The male palmate also has two, slightly raised, dorsal ridges, one on either side of the breeding crest, giving them a squared off or boxy appearance. The male's underbelly is usually off white, with a light yellow central stripe and lightly mottled with small dark irregular spots.

39: April, a male palmate newt in the pond, approx 6cm in length.
The breeding crest, dorsal ridges and black webbed rear feet are clearly visible.

Skin colour among individual males ranges from almost black through to olive green, with the flanks, head and front legs heavily speckled with small, light spots. On entering the pond and coming into breeding condition, the male's tail becomes flared. Uniquely among our native newts, the male palmate newts tail comes to an abrupt, truncated end, from which emerges a hair thin, black filament about five millimetres long. Because the males lack a clearly visible breeding crest, differentiating between male and female palmates in the pond is more difficult than with other newts.

Females in the pond

Female palmate newts start to return to the breeding ponds in March, as with the males the return of individual newts is normally spread over a six-week period. The females return to the pond already laden with the season's eggs. They can be identified in the water by their egg swollen bellies and the lack of a breeding crest or dorsal ridges. Skin colour ranges among individual females from light brown to almost black with mottling over the head, body and tail. The female palmate newt's hind feet, unlike the male's, are not webbed or coloured black, the tail does not flare and tapers to a sharp point without the protruding, black filament.

38: April, a female palmate newt in the pond, approx 6cm in length.
This photograph shows the female's egg swollen appearance, mottled skin pattern, lack of a breeding crest, absence of any dorsal ridges and the un-webbed rear feet. The underbelly is off white and speckled with small, dark, irregular spots, with a light yellow central stripe

Correct identification of female palmate newts can be problematic. This is because female palmate and female smooth newts are very similar to each other in size and appearance. At ponds known to support breeding colonies of both species it can be very difficult for the inexperienced eye to differentiate between the females.

One method regularly used by ecologists to determine the species of a captured female newt, is to look at the underside of her chin. Smooth newt females normally have a spotted or freckled chin while the female palmate's chin is usually unspotted.

39: May, a female palmate (left) and female smooth (right).
These two female newts clearly show the similarities and subtle differences between the two species and the potential difficulties of correct identification.

40: Female palmate with unspotted chin.    41: Female smooth with a spotted chin.

As the breeding season comes to an end the adult palmates undergo a number of physical changes in preparation for leaving the pond. The male palmates absorb their breeding crests, the dorsal ridges become less pronounced, the flared tail thins and the hind feet become lighter in colour though the webbing between the rear toes remains.

The female palmates become much sleeker as egg laying comes to an end. After leaving the pond the skin of both sexes becomes dry and leathery in texture and appearance and its ability to absorb oxygen from water is lost. During their terrestrial phase male and female palmates closely resemble each other, and identifying the species and sexing adult newts can be very difficult.

Adult palmate newts in their terrestrial phase

After leaving the pond, at the end of the breeding season, and entering their terrestrial phase the male palmate newt's key identifying features are; the hind feet, which remain visibly webbed throughout the year and the black filament, at the end of the truncated tail. This tail filament is still visible in the terrestrial phase but quite hard to see, as it shrinks down to barely one millimetre in length after the males leave the water.

42: August, a male palmate in his terrestrial phase, approx 6cm in length.
Identified as a male palmate by the webbed toes on the rear feet and a truncated tail with a just visible remnant of the black tail filament, because of similarities in size and appearance male palmate newts are often confused with male smooth newts.

The male palmate newt's body colour and markings vary considerably from one individual to the next. Confusingly, when found away from the pond in their terrestrial phase, male palmates look very similar to adult smooth newts. Only a careful examination, to check for the distinctive webbing on the newts' hind feet and the remnant tail filament, will ensure that you correctly identify any terrestrial phase adult palmate newts that you find in your garden.

Their dry, leathery skin (in the terrestrial phase) enables them to roam over quite a wide area in search of food and shelter without drying out. This explains why they are sometimes found in gardens without ponds. In common with smooth newts they are occasionally mistaken for lizards, though like all newts they will 'freeze' and allow you to pick them up if you disturb them sheltering under stones or logs during the day.

After leaving the pond the female palmate's key identifying features are; a uniform dull brown coloration, lack of webbing between the toes on the rear feet, the narrow band of pale yellow running along the bottom edge of the tail and the translucent, off white unspotted chin.

43: September, a female palmate newt in her terrestrial phase, approx 7cm in length.
Identified as a female by her uniform brown skin colour, unspotted chin, the lack of webbing on the rear toes and the tail which comes to a sharp point and lacks the tail filament present in the male.

From July to November the adult palmate newts are in their terrestrial phase living and feeding in and around the garden. During the day they take cover in a variety of summer shelters, emerging after dark or rain to hunt for earthworms and insects.

As with the other newt species, success or failure in hunting and finding shelter in their terrestrial phase impacts directly on their breeding success the following spring. From late October the adult palmates instinctively seek out frost-free winter shelters. Locations include compost heaps, log piles old walling and mammal burrows, often several individuals will find and share the same winter shelter.

As the days shorten and the temperature falls, the newts stop feeding; with the onset of winter and the first frosts they enter a torpid state and become immobile. Care should be taken not to disturb them during this period as exposure to extreme cold or a hard frost can kill them. The following year, with the arrival of spring and warmer weather, the adult palmate newts emerge from their winter shelters and begin to make their way back to the breeding ponds.

Palmate newt breeding behaviour

Adult palmate newts, of both sexes, begin the journey back to the breeding ponds from their winter shelters during March and April, usually at night and after rain, when the temperature is above 10°C. Cold weather and chilling winds brings the return to a temporary stop with the newts taking shelter, until it warms up again. Breeding begins as the water temperature starts to rise in March and continues through to June. Courtship and mating can only take place in water and follows a similar pattern to the great crested and smooth newt.

The stages of palmate newt courtship and mating;

Chase and Scent
The male palmate chases after the female and scents her cloaca to confirm that she is the right species, sex and receptive to his advances.

Display and Fan
The male positions himself in front of the female and displays by arching and twisting. At the same time he uses his tail to fan scents and pheromones from his swollen cloaca towards her as he slowly backs away. To keep her interest, during calmer periods of the tail fanning stage, a male will often gently wiggle the black filament at the end of his tail under the female's nose.

Following
When suitably impressed the female starts to follow the male around the pond. The male continues to display, fan and wiggle his tail filament.

The Spermatophore
After several minutes of following, the male turns to face away from the female, raises his tail and releases a sticky packet of sperm called a spermatophore from his cloaca. He then turns back to face the female and continues to display and fan leading her forward.

Fertilization
As the female passes over the spermatophore it attaches to her cloaca. The spermatophore is taken up inside her body, via the cloaca. The sperm is then released and fertilizes a portion of her eggs.

Palmate newt courtship displays are no less vigorous than those performed by smooth newts, but lacking a substantial breeding crest, the males use their black, webbed, hind feet in the courtship display. By spreading their toes, the webbing is extended and the feet resemble black paddles signalling to the female the male's suitability as a mate.

The difference in scents and pheromones released from both the male and female cloaca combined with variations in their courtship displays is usually enough to prevent our native newt species mating with each other. However it is known that under certain conditions smooth and palmate newts will cross breed, especially when kept together in captivity. The viability and success of the hybrid offspring resulting from these matings is little researched and the incidence of cross breeding in the wild and its effect, if any, on the general population of both species is currently unknown.

As with the other newt species, individual palmates, of both sexes, will mate with multiple partners over the breeding season and not all spermatophore transfers are successful. Following her first successful mating the female palmate begins to lay her eggs

Palmate newt eggs

Egg laying, via the female palmate's cloaca, starts within a few days of their first successful mating and reaches a peak during April and May. The females attach each sticky-coated egg individually onto the leaf of a suitable water plant and then use their hind feet to fold the leaf protectively around each one. Most eggs are laid during the day in submerged vegetation just below the pond surface, female palmate newts will take several minutes to carefully lay and wrap each egg. The larger females can lay up to one hundred and fifty eggs over a season and will spend up to twelve weeks in the pond laying them all.

Palmate newt eggs are oval and approximately two millimetres in diameter with a grey brown nucleus enclosed in a sticky, clear jelly shell. In the field it is not possible to differentiate between smooth and palmate newt eggs as they are very similar in both size and appearance. Research suggests that, when given a choice, the female palmate newt prefers to lay onto smaller leaved aquatic vegetation.

44: April, several palmate newt eggs, laid onto flote grass.
Palmate newt eggs are approx 2mm in diameter, with a brown nucleus inside a clear, jelly shell. Some of the flote grass blades are still folded others have been carefully teased open to reveal the eggs.

By folding the leaf around the egg the female palmate newt hopes to hide and protect it from predators. It is not unusual for the same plant to be used repeatedly by one or more females, resulting in water plants and pondweeds with numerous multi-folded leaves. Due to the prolonged spawning season it is quite normal to find palmate newt eggs at various stages of development in the pond, often on the same plant, at the same time.

Palmate newt egg development

As with the other newt species the rate of palmate newt egg development is governed by temperature, with warm weather speeding it up and cold slowing it down. Eggs laid in March can take up to four weeks to develop into tadpoles, while eggs produced later in the season, when the water is relatively warmer, only take about two weeks to become free swimming tadpoles.

The palmate egg nucleus starts life as a small, brown oval, inside a clear, jelly shell. As the cells in the nucleus divide it gradually elongates forming the head, body and tail of an embryonic palmate newt tadpole. Continuing to grow, but constrained by the jelly shell, the developing tadpole gradually curls itself round inside the egg, forming itself into a circle, with the head touching the tip of the tail.

45: May, four palmate newt eggs, approx 2mm in diameter.
The blades of flote grass the eggs were laid on have been carefully teased open to show the development of the palmate newt egg nucleus into an embryonic tadpole.

Shortly before palmate tadpoles emerge from their eggs, dark camouflage stripes form along the length of their head, body and tail. After vigorously twisting and turning inside the egg they eventually break the jelly shell open and swim off into the pond.

Given warm weather and a good supply of microscopic prey the palmate tadpoles will grow and develop very quickly. It normally takes them about sixteen weeks to complete their metamorphosis from water living tadpoles into land-based young newts.

Palmate newt tadpole development

Week one

The newly emerged palmate newt tadpoles are very small, only two to three millimetres in length. They have fully formed eyes but no legs and breathe using a set of external gills situated on either side of the head. At this age and stage of development they closely resemble great crested and smooth newt tadpoles; this makes the correct identification of very young smooth and palmate newt tadpoles' virtually impossible in the field. The newly emerged palmate tadpoles are very delicate and will quickly die if removed from the water or injured.

46: May, a newly emerged palmate newt tadpole, approx 2.5mm in length.
At this age the tadpoles of all our native newts are very similar in appearance and difficult to tell apart, this is especially true of smooth and palmate newt tadpoles. The five pence piece in the background has been included to give an idea of size.

Immediately after breaking free of the egg, the palmate newt tadpoles seek shelter among the pondweed and water plants. There they rest for several hours using up the last reserves from the egg nucleus, before starting to hunt for microscopic invertebrates and insect larvae, which are caught live and swallowed whole.

During these first weeks of life palmate tadpoles rely on their camouflage coloration and hiding among the pondweed to protect them from hungry predators. Even so large numbers fall prey to a host of pond predators, this predation continues unabated until they leave the pond as young newts.

Week four

At four weeks old palmate newt tadpoles are about one centimetre in length and have grown their front legs complete with toes. Their feathery external gills, used to absorb oxygen directly from the water, are now clearly visible. The tadpole's large eyes are used to search the pond for tiny prey items including water fleas and mosquito larvae that are caught with a quick snap of the mouth and swallowed whole. To avoid predators the tadpoles instinctively avoid open water and keep to the weedy areas of the pond where they feed continuously and voraciously. With a good supply of food and warm weather they continue to grow and develop rapidly.

47: June, a young palmate tadpole about four weeks old and approx 1cm in length.
The front legs are always the first to grow in the tadpoles of all three of our native newt species.
All newt tadpoles have the ability to grow new limbs if they are lost or damaged.

Although extremely fragile and delicate, the palmate newt tadpole's newly developed front legs are strong enough to help them clamber through the pondweed in search of prey. By about six weeks old their back legs will have started to grow.

Amazingly, and in common with all our newt species, if a palmate newt tadpole loses part of a limb, to either a predator or accident, it has the ability to grow a new toe, foot or entire limb. This unusual ability is the subject of ongoing medical research, looking at the possibility of regenerating lost or damaged human organs.

Week eight

At eight weeks old most palmate newt tadpoles are about two centimetres in length and have grown their back legs. Skin colour among individuals can range from light through to very dark brown. The tadpoles continue to keep to the weedy areas of the pond searching for prey and hiding from predators. At this age they have a silvery white underbelly tinged with pink; their external gills are well developed, clearly visible and still used to take some oxygen directly from the water. However, like all newt tadpoles, they must regularly rise to the surface to take a gulp of air or they will drown. As they grow the range and size of prey items increases, and In common with all newt species it is not unusual for the larger palmate tadpoles to attack and eat their smaller siblings.

48: July, a palmate newt tadpole about eight weeks old and approx 2cm in length.
With four legs the palmate newt tadpole is starting to look more like a newt.

From September onwards, in common with the other newt species, any palmate tadpoles remaining in the pond, (resulting from eggs laid late in the season), will suspend their development. These tadpoles will spend the winter months in the pond over wintering if they survive the winter they will resume their development the following spring as the weather warms.

This over wintering strategy ensures that no eggs or tadpoles are wasted and gives those that survive the winter a head start on the tadpoles resulting from the next year's spawning. The survivors will leave the pond in early summer of the following year and join their siblings in and around the garden hunting for insects and invertebrates.

Week twelve

At twelve weeks old palmate newt tadpoles are about three centimetres in length and their external gills have started to be absorbed into the body. Their skin has thickened and darkened in preparation for the move out of the pond onto land. At this stage of their development the tadpoles start congregating at the surface and in the weedy shallows around the edges of the pond.

49: July, a palmate newt tadpole about twelve weeks old and approx 3cm in length.
The tadpole's skin has started to thicken, the external gills have started to shrink and it spends increasing longer periods resting at the surface and in the shallows. Skin coloration among individual palmate newt tadpoles varies considerably ranging from dark through to light brown.

At twelve to fourteen weeks old palmate newt tadpoles will occasionally leave the water and venture into the emergence zone where they rest for a few minutes, before returning to the pond. This behaviour continues for a few days until the tadpoles have completely absorbed their external gills. Once the gills have been lost they cannot be re-grown and the tadpoles will leave the pond as young palmate newts. As young newts they must have easy access to the emergence zone around the pond, otherwise there is a very real risk of them drowning.

Palmate and smooth newt tadpoles are particularly hard to distinguish from each other, which can be problematic as the two species will often share the same ponds. As a general guide palmate newt tadpoles are slightly smaller than same age smooth newt tadpoles and have a pink tinged underbelly,

The difficulty of correct tadpole identification (palmate/smooth) is such that even experienced ecologists and amphibian workers, will make mistakes when attempting to identify individual tadpoles from ponds known to contain breeding colonies of both species.

Week sixteen

At sixteen weeks old and about three centimetres in length most palmate newt tadpoles have left the pond and become young palmate newts. After spending a few of days resting and gathering their strength hidden in the emergence zone they move out into the garden in search of food and shelter. Young palmate newts will not return to water until they are ready to breed, usually at three years old. During these land-based years a few young newts may travel several kilometres. If they are not lost to predators or accident, they might, given a little luck, establish new breeding colonies in neighbouring ponds. Most young palmate newts are much less adventurous and spend their first three years living within a two hundred-metre radius of their birth pond.

50: August, a young palmate newt about sixteen weeks old and approx 3cm in length.
Newly emerged from the pond this young palmate newt will spend a few days in the emergence zone before seeking food and shelter in the surrounding gardens. The penny in the background gives an idea of size.

Young palmate newts spend their days hidden away in summer shelters around the garden, emerging at night or after rain, to hunt for small insects and invertebrates including aphids and very small worms. As the weather starts to cool in late October they instinctively seek out a safe, frost-free winter shelter. With the onset of winter and the first hard frosts they stop feeding and enter an immobile, torpid state. As the weather warms, the following spring, they emerge with the young from previous years and resume hunting for prey.

All our native newt species are capable of colonising urban gardens, however, a natural colonisation can only occur if they have a safe, navigable route, from an existing population into a suitable site. In many areas, because of roads, housing and commercial development these green corridors no longer exist; consequently the natural dispersal of newts, in some areas has been severely impacted.

Chapter six

Establishing newts in your pond

General Information

As a general rule, if newts are already present in your area, there is an excellent chance that within four years they will have established themselves naturally in any new pond that provides them with a suitable aquatic habitat. This occurs as part of their normal dispersal and colonisation behaviour and simply requires that you be patient.

Assuming conditions in your pond are suitable the first adult newts could start to arrive in the spring, of the year following its creation. In order to confirm their presence you should consider undertaking regular searches of the pond from March through to May, looking for the adults and most importantly their eggs.

If they fail to arrive by the spring of the fourth year, it is either because there are no newts living within a five hundred-metre radius of your pond, the area lacks the green corridors necessary for adult and young newts to successfully disperse from neighbouring colonies and reach new sites, or conditions in your pond and garden are unsuitable.

It is possible to introduce newts into new ponds. However such introductions, even the best intentioned and carefully planned, are fraught with difficulties. They often prove unsuccessful and can result in the death of the introduced adult newts or eggs.

I would strongly recommend that you resist the temptation to introduce newts into your garden pond and allow nature to take its course. If they establish themselves naturally in your pond and garden you can be sure that the conditions and habitat are suitable for them.

However, recognising that some garden pond newt introductions are likely to occur regardless of my advice, I have decided to include two methods of introducing smooth and palmate newts into new garden ponds in the hope of preventing unnecessary cruelty to the animals.

The introduction of smooth or palmate newts should only be undertaken if:

- You are absolutely certain that conditions in your pond and garden are suitable.
- Smooth or palmate newts have no means of reaching your garden naturally.
- You seek further advice from environmental organisations before attempting to introduce them.

The two methods of introducing smooth or palmate newts into a new pond, described in this book, are dependent on:

- Finding a suitable garden pond in your locality with a co-operative owner willing to help.
- The donor pond supporting a healthy, verified, breeding population of at least one hundred positively identified adult smooth or palmate newts.
- Having the ability to positively identify our three newt species and their eggs.

Under no circumstances should any adult newts, their eggs or young be taken from the wild.

Great crested newts, their eggs or young should never be taken from any pond.

1: Adult newt transfer

In certain circumstances adult newt transfer can be used to attempt the establishment of a smooth or palmate newt colony in a new pond. In simple terms it involves the physical capture and transfer of a small number of adult smooth or palmate newts from an existing garden pond in your locality. Ideally newt capture and transfer should be undertaken by somebody with relevant experience.

Moving adult smooth or palmate newts

- Capture and transfer of adult smooth and palmate newts should be undertaken as soon as possible after their return to the donor pond in early spring.

- March is probably the best month to move adult newts to a new pond, preferably during cold weather when the temperature is around 5°C.

- Moving adult newts to a new pond in early spring, especially during cold weather, appears to fool them into believing that they are still in the donor pond, increasing the likelihood of them remaining in the new pond to breed and maximises the number of eggs they produce.

- Mild weather transfers should be avoided. They are usually less successful and can fail completely because the adult newts, acting on their strong natural homing instinct, abandon the new pond and attempt to find their way back to the donor pond. This instinct driven attempt to get 'home' may result in their death through predation or accident.

- To minimise the risk of injury adult newts should be transported out of water in a secure, well ventilated container, lined with a thick layer of damp moss.

- On arrival at the new pond the adult newts should be carefully released into dense aquatic vegetation around the waters edge.

- Depending on the success of the transfer and the number of adult newts introduced into the new pond, hundreds or possibly even thousands of newt eggs will be laid over the breeding season.

- Given good aquatic habitat, the survivors from this spawning will leave the new pond about sixteen weeks later as young newts, and should set up home in the garden and its surrounds. These young newts, having naturally imprinted on the new pond, will, given good terrestrial habitat, return at three years old to breed.

- Some of the transferred adult newts may take up residence in the garden and its surrounds after they leave the pond at the end of the breeding season. These newly 'resident' adults may return to the new pond and breed again in subsequent years.

- It is likely that the majority of transferred adults, acting on their strong homing instincts, will attempt the hazardous journey back to the donor pond the following spring.

Further information and advice about the advisability of moving adult newts can be obtained from your Wildlife Trust, local ARGs, Froglife, Natural England, Countryside Council for Wales and Scottish Natural Heritage or at www.newtsinyourpond.com (see useful contacts page 89).

2: Newt egg transfer

Newt egg transfer is recognised by some conservation organisations as an acceptable way of establishing smooth or palmate newts in a new pond if;

- They fail to arrive naturally.
- The new pond is inaccessible to neighbouring newt colonies.
- Conditions in the pond and garden are suitable.
- The newt eggs are positively identified as smooth or palmate.

This method of establishing newts in a new pond minimises any adverse impact on the donor pond's newt population and reduces the risk of transferring amphibian and fish diseases. The process is relatively simple and involves collecting newt eggs from the donor pond during the breeding season.

To make the egg collection and transfer process less laborious, it can be a good idea to place a few hand sized, fan shaped bundles of flote grass into the donor pond just below the waters surface. With a little luck the female newts will use these grass bundles for egg laying making it easy to collect the eggs and transfer them to the new pond.

Newt eggs should be transferred to the new pond attached to the grass blades or pondweed in a water filled container, with care taken to ensure that they remain immersed in water. If the eggs dry out the developing embryos will die.

To give the newt eggs and the resulting tadpoles the best possible chance, the transferred grass blades or pondweed with eggs attached, should be carefully placed just below the waters surface among submerged pondweed in a sunny area of the new pond. The pond should be regularly checked to ensure that any falls in water level (due to evaporation), does not leave the eggs exposed above the water. When the eggs hatch, the newt tadpoles will continue their development as if they had never been moved.

When the survivors leave the water as young newts in late summer, they will naturally imprint on the new pond and the majority should establish themselves within a two hundred metre radius of what they believe to be their birth pond. After three years they will return as adults to breed in the new pond. One successful transfer of newt eggs can be enough to establish a new colony, though to improve the chances of success the process should be repeated over three breeding seasons.

Surprisingly accidental newt egg transfer is fairly common and, after natural colonisation, it is believed to be the way many new garden pond colonies become established. Garden centers and specialist water plant suppliers play a part in this accidental newt egg transfer, as many are known to have newts living and breeding in their display ponds and holding tanks. Over the years many unsuspecting members of the public have carried home a bag or two of oxygenating pondweed from the local garden center with newt eggs hidden away in leaf folds. Three years later, usually much to their surprise and delight, these lucky people have found adult newts breeding in their garden pond with, until now, no idea where they came from.

Another accidental source of newt eggs can be well meaning, garden pond owners who generously give pondweed and aquatic plants, with unbeknownst to them newt eggs attached, to friends and family to help them establish a new garden pond or water feature. While you are unlikely to do much harm if you transfer pondweed with some positively identified smooth or palmate newt eggs attached, the moving of any great crested newt eggs could result in a large fine and or a period of imprisonment!

Advantages and Disadvantages of transferring adult newts or their eggs to establish new newt colonies:

Advantages:

- It can work very well if done properly by experienced amphibian workers with a good understanding of the process and the newts' natural behaviors, habitat requirements and life cycle.

- Transferring adult newts or their eggs overcomes the problems associated with the lack of green corridors, allowing them to colonise sites they could never hope to reach naturally.

- It helps to expand their range and distribution; in addition newt colonies established using transferred newts or eggs can quickly become a source of colonisation as the population grows.

- Newt colonies created in this way can act as replacements for those lost when garden ponds are destroyed or fish are introduced into established breeding sites.

- Transferring adult newts can help improve the gene pool of isolated colonies and prevent problems including localised extinction caused by isolation and inbreeding.

- Importantly, can secure the positive, ongoing involvement and interest of new pond owners, who having constructed a pond for newts, are keen to have them arrive as soon as possible.

Disadvantages:

- It is not always successful, despite using the best methods, and can result in the death of some or all of the transferred adult newts or eggs.

- A failure to ensure that the donor pond newt population is large enough can impact adversely on its continued viability, possibly leading to a localised extinction.

- It may upset the ecosystems of older established ponds. This can occur as a consequence of transferred newts at one site, colonising neighbouring ponds and destabilising existing amphibian and invertebrate populations.

- There is a risk of transferred newts carrying and spreading amphibian and fish diseases.

Moving any amphibians between garden pond sites is a contentious subject with valid arguments on both sides. It should only be considered as a last resort, after much thought. Further information and advice can be obtained from your Wildlife Trust, local ARGs, Froglife, Natural England, Countryside Council for Wales and Scottish Natural Heritage or at www.newtsinyourpond.com

Once newts have successfully established themselves in your pond and garden it can be fun, educational and potentially important to keep a record of the colony. By monitoring and recording your newts, you will greatly increase your knowledge and understanding of them. For some children keeping a newt diary or a log book of newt-related events in the pond and garden can be the start of a lifetime's interest in nature and conservation.

Chapter seven

Monitoring and Recording

General information

The aim of monitoring and recording is to build up a historical record of your newt colony and help you to become familiar with the newts living in your pond and their behaviour. This chapter offers suggestions on record keeping and things to look out for when observing newts in the pond and garden over the course of the 'newt year'.

Monitoring of your newt colony should start in February and go on through to November. To produce useful records it needs to be done on a regular basis either weekly or monthly and should include day and night time observations of both the pond and garden in various weather conditions.

The degree of detail in your records is entirely up to you and will depend on your level of interest and enthusiasm. Records can productively range from basic notes on when the first adult newts return to breed each year, to more complex records on population size, survival rates, life spans, dispersal patterns and observed behaviour.

Observing newts in the pond during the day is simply a case of spending time sitting or standing quietly by the side of your pond and looking into the water and watching them. These observations can be supplemented and enhanced using a hand net, a "dipping bowl" and a tank. Try to limit the use of the hand net when catching newts for identification and study, as it will cause disturbance and damage to habitats by stirring up bottom sediments and uprooting pondweed which may contain eggs.

The easiest and least disruptive way to collect specimens from the pond is to use a large white plastic bowl dipped quickly into the water. The creatures swept into the bowl by the inrush of water can then either be identified in the bowl and returned to the pond, or carefully transferred to the tank for further study. This "dipping bowl" method is a particularly good way to check on the progress of newt tadpoles without injuring them.

Observing newts at night is known as torching, it involves using a strong torch to search for them either in the pond or around the garden and spending time watching their nocturnal activities. Mild rainy nights in spring, summer and autumn about an hour after dark are best, although most nights will usually reveal something of interest.

To prevent undue disturbance when torching at the pond, keep the torch steady and move the beam slowly, if disturbed, the newts will take cover in deeper water, if this happens turn the torch off and wait a few minutes before trying again. It is a good idea to let your neighbours know that you will be torching, in order to prevent calls to the police reporting prowlers in the area! If you positively identify great crested newts, in your pond or garden, you should stop torching, (it is an offence to disturb them) and restrict future monitoring to daytime observations.

To give you a few ideas and help get you started I have included in this chapter a few of my own records, a quick guide to the newt year that details times and things to look out for, and a basic amphibian recording sheet. You may have to adjust the timings slightly to accurately reflect what happens in your pond and garden.

Sample record one

Please note the author is licensed by Natural England to survey for great crested newts.

Date
25th December 2003

Time
10.30pm

Weather
Cloudy, mild, light drizzle

Temperature
12°c

Location
Macclesfield - Garden Pond

Type of search
Torch

Duration
Forty-five minutes.

Species found
Great crested newt

Sex
Female

Numbers
1

Activity observed
A female great crested, about 17cm in length and heavily swollen with eggs, was seen in the pond moving slowly through the starwort (pondweed) about 15cm down, disturbed by the light I lost sight of her when she swam off into deeper water.

Comments

What a surprise! I had an idea that they might be on the move because of the recent mild weather but I never expected to find them already in the pond. This female newt has actually returned to the pond twice within the same year. Could this be an indication of climate change? No other newts were seen in the pond or elsewhere in the garden.

Sample record two

Date
9th May 2004

Time
10pm

Weather
Warm, dry.

Temperature
11°c

Location
Macclesfield – Garden Pond

Type of search
Torch

Duration
One hour

Species found
Great crested newts, Smooth newts

Sex
Great crested – approx eleven females, nine males
Smooth - approx thirty four females, thirty seven males

Numbers
Great crested **Total 20**
Smooth **Total 70**

Activity observed
I observed about eight adult smooths and a couple of great crested newts at the surface.

Comments

The purpose of torching was to assess the size and health of the newt populations and compare the results with the last count done in May 1998. The results show a small rise in the populations of both species and roughly equal numbers of both sexes.

Results of the 1998 count.
Great crested - females nine, males seven. **Total 16**
Smooth - females thirty-three, males thirty-one. **Total 64**

Sample record three

Date
15th May 2006

Time
10pm

Weather
Warm, dry no wind

Temperature
14°c

Location
Macclesfield - Garden Pond

Type of search
Torch

Duration
One hour.

Species found
Great crested and Smooth newts

Sex
Great Crested – approx eighteen females, twelve males
Smooth - approx ninety five females, eighty seven males

Numbers
Great Crested **Total 30**
Smooth **Total 182**

Activity observed
Surfacing adults were observed.

Comments

I believe the recent run of mild winters has increased survival rates among the adults and young of both newt species breeding in my pond. This appears to have resulted in a substantial increase in the populations of both species.

The downside has been the complete loss of all this season's frogspawn. It is likely that the pond and surrounding habitat will not be able to support newt populations at these levels. I anticipate that increased predation of their own young, cannibalism and a sharp increase in the number of other pond predators fuelled by the increase in newt eggs and tadpoles will bring the populations back into balance over three to four years, as could a run of hard winters.

A quick guide to the newt year

January

Adults and young newts are under cover in winter shelters around the garden. A few newt tadpoles from last year's spawning are still in the pond, having suspended their development for the winter. They retain the external gills allowing them to take dissolved oxygen directly from the water. Try dipping for these over wintering tadpoles.

February

The first adult newts return to the pond towards the end of February, earlier if the weather is particularly mild. Try torching for them as they cross the garden or in the pond. Record the date you see the first adults, include the sex and species.

March

Adults continue to arrive at the pond throughout March. Regular torching and night searches should allow you to get a good estimate of the population's size.

Adult newts come into breeding condition, courtship displays and mating begins. The females lay their first eggs. Start checking the pondweed for folded leaves containing newt eggs and record the date you find the first one and if possible the species.

As the weather warms and the days lengthen, the over wintering newt tadpoles from last year's spawning resume their suspended development.

April

Late returning adults arrive back at the pond. On warm sunny days you can expect to see adult newts basking at the water's surface with their backs just out of the water, if disturbed they will dive to the bottom in a flurry of splashing tails, returning to the surface when the danger has passed.

During the day male smooth and palmate newts will be seen chasing the females and performing their courtship displays, females may be seen among the pondweed as they lay and wrap their eggs.

Adult newts can be observed rising up through the water as they swim to the surface to take a gulp of air. In open water some adults may be seen floating in the water column feeding on clouds of water fleas, taking them with quick jerky motions and a snap of the mouth or chasing after frog tadpoles. Try dropping some chopped earthworms into the pond and watch the resulting feeding frenzy.

The first of the newt eggs laid in March are now hatching, try catching the newly emerged tadpoles using a "dipping bowl" and record the date you find the first one.

May

By May most of the adult newts in the colony have returned to the pond and egg production reaches its peak. Tadpoles from the first eggs laid in March have grown their front legs; check on their progress using the "dipping bowl". The over wintering tadpoles from last years spawning start to leave the pond as young newts. They take up residence in the garden alongside previous seasons young.

June

The first adult newts, those that returned in late February early March, begin to leave the pond, eggs continue to be laid by the late returning females. The oldest newt tadpoles have grown their back legs, check on their progress using the "dipping bowl".

Begin searching summer shelters around the garden, you are looking for terrestrial phase adults and the previous seasons young, record the species, sex and numbers found, include the time, date and weather conditions. Try leaving an old piece of carpet in a neglected corner and check under it every few days for adult and young newts using it as a summer shelter.

July

The majority of adult newts have left the pond. The first of the season's tadpoles begin to leave the pond as young newts. The last eggs are laid, the resulting tadpoles, lacking sufficient time to develop into young newts will have to overwinter in the pond. Continue checking on newt tadpole development using the 'dipping bowl'.

August

The last of the adults and the majority of the season's young newts leave the pond. They spend the days under cover and the nights hunting around the garden. Continue checking summer shelters.

September

The last of the season's young newts leave the pond. As the water temperature falls and the days shorten the small number of newt tadpoles still left in the water stop developing and prepare to spend the winter in the pond.

October

The adults and young spend the days under cover and the nights hunting for live prey. Towards the end of October they start to seek out winter shelters. Record the date you stop finding them in their usual summer shelters.

November/December

Adult and young newts are under cover, hidden away in winter shelters around the garden, a small number of over wintering newt tadpoles are still in the pond.

Adult and young newts should not be disturbed in their winter shelters, exposure to frost can kill them.

Monitoring; continue with occasional torch searches when the weather is favorable.

When you can expect to find newts in your pond

- Dark blue represents maximum numbers in the pond.
- Light blue indicates possible presence.

Newts In the Pond	Jan	Feb	March	April	May	June	July	Aug	Sept	Oct	Nov	Dec
Adults												
Eggs												
Tadpoles												

Basic Amphibian Recording Sheet

Key

M = Male
F = Female

S = Smooth newt
P = Palmate newt
GCN = Great Crested Newt

CF = Common Frog
CT = Common Toad

Date	Location: Grid Ref if possible	Frog species	Toad species	Newt species	Weather temp	Eggs /spawn breeding
8/3/06	Macclesfield Garden pond	50 + M & F **CF**	0	10M & 5F **S** 3M & 2F **GCN**	Sunny 14°C	**CF** spawn **S** & **GCN** eggs

Chapter eight

Observing and handling newts

In the pond and garden

Adult smooth or palmate newts caught in the pond should be studied in a tank or similar water filled container, as many male newt characteristics are most apparent when seen under water. These characteristics include the male smooth newts' breeding crests, flared tails, and the male palmate newt's distinctive, black, webbed toes on the rear feet. Healthy, pond-caught newts will be active, bright eyed, plump and depending on the species and sex have striking body markings.

Smooth and palmate newts found around the garden can be hand caught and safely kept in a damp, moss lined, secure and well-ventilated container for up to one hour, while you establish their species, sex and numbers. Daylight searches of summer shelters in your garden during August and September will give you a good idea of newt numbers, their condition and breeding success. A healthy well-fed terrestrial phase newt will be plump, bright eyed and, after a brief period of immobility when first captured, it should become quite active.

51: September, fourteen smooth newts found in the flower border.
Eight males, one female and five young newts from sixteen weeks to two years old.

Don't be afraid of handling adult and young smooth and palmate newts whether caught in the pond or garden, they will come to no harm if held carefully for short periods, though to reduce stress and prevent damage to their skin you should always wet your hands first. Often a close, hand held examination is the only way to correctly identify the species and the sex of individual newts.

Keeping adult newts in a tank

Providing they settle quickly, keeping four adult smooth or palmate newts from your pond, in a good sized, covered tank for a few days will cause them no harm and give you the opportunity to observe behaviours that might otherwise be difficult to see in the pond. These behaviours include the intricacies of adult courtship, spermatophore transfer, egg laying, leaf folding and feeding. Spending time observing them in a tank will help you to recognise the same behaviours in the pond.

Looking after adult smooth and palmate newts in captivity involves setting up a tank to resemble the breeding pond, occasionally changing the water to maintain its quality and providing them with live food daily. Adult newts, though skilled escapologists, are relatively easy to keep in a tank and will happily feed on small pieces of freshly chopped earthworm, frog tadpoles and small invertebrates brought in from the garden pond.

52: March, a male smooth newt, in a tank set up to resemble the breeding pond.
Remember it is an offence to keep adult great crested newts in captivity.

Adult smooth and palmate newts can be caught and moved into a tank at anytime between March and May, though early transferred newts are most likely to accept the change. If they refuse to feed, start repeatedly swimming into the corners or are continually trying to escape, you should release them back into the pond. By late June, as they start to enter their terrestrial phase, any adult newts still in the tank must be released at their place of capture and allowed to resume their natural behaviour.

Observing and handling newt eggs

When gently unfolding or teasing apart the leaves of pondweed or flote grass to expose newt eggs for identification, you must take great care not to crush, split or tear the eggs jelly shell, any damage will result in the destruction of the developing nucleus.

Examination of newt eggs, using a magnifying glass to reveal detail, can be undertaken out of the water for short periods (two minutes), without harming them. Newt eggs exposed for examination are likely to be taken by predators when they are returned to the pond, for this reason you should only unwrap a small number of leaves containing eggs in order to confirm a breeding newt presence, and if possible the species. Remember you need a licence to disturb great crested newt eggs.

53: April, blades of flote grass teased open to reveal the eggs of all three native newt species. A white great crested egg can be seen in the centre of the photograph. Directly above it is a grey smooth newt egg. To its right a slightly smaller, brown coloured palmate newt egg. Great crested newt eggs can only be disturbed under licence from Natural England or the relevant agencies in Scotland and Wales.

Bringing a few smooth or palmate newt eggs indoors and placing them in a bowl of pond water, on a window sill, out of direct sunlight, will speed the eggs development and allow you to observe their transformation into newt tadpoles close up. Occasionally newt eggs will be attacked by fungal growths and viral or bacterial infections which destroy the developing nucleus. Any newt eggs that take on a cloudy or milky appearance should be removed to prevent the fungal growth or infection spreading to the other newt eggs in the bowl.

54: May, great crested egg left and smooth newt egg right, approximately 5mm and 3mm in diameter. Both of these newt eggs were laid at about the same time and are on the verge of hatching.

In the above photograph the great crested newt egg has become detached from its protective leaf. In the pond these exposed eggs and their developing embryos would quickly become a meal for one of the pond's many predators. Already there is a clear difference in size between the tadpoles of the two species. The developing tadpoles' camouflage body stripes and fully formed eyes are clearly visible through the eggs clear jelly coating.

When the tadpoles emerge it is always advisable to release them back into the pond, as they require immediate access to large quantities of microscopic prey. If you are prepared to take the time to look after them properly you can continue to follow their development, by carefully transferring a small number of newt tadpoles into a tank filled with pond water and weed.

Keeping newt tadpoles in a tank

Keeping a small number of smooth or palmate newt tadpoles in a tank for a few weeks should cause them no harm, and will give you the opportunity to observe their development into young newts first hand. Looking after them is a time consuming business and involves setting up a tank to resemble a natural, weedy pond, occasionally changing up to a third of the water with rain or pond water to maintain its quality, and providing the newt tadpoles with a daily supply of live invertebrates and insect larvae direct from the garden pond (water fleas are ideal for young tadpoles). As the tadpoles grow so does the size of the prey items that they will take.

Great care must be taken not to accidentally introduce predatory water beetles, larvae and nymphs into the tank when setting it up, changing the water or feeding, otherwise all your efforts will be wasted and you will look into the tank one morning and find that all the newt tadpoles have been eaten!

55: July, a smooth newt tadpole and a same age great crested newt tadpole.
The smooth newt tadpole is in real danger of becoming a meal for its bigger cousin.

To observe their development, smooth and palmate newt tadpoles can be taken from the pond and transferred to a tank at any time from April through to early August using a "dipping bowl". The later you collect newt tadpoles the more developed and robust they will be.

It is very important that all the newt tadpoles kept together in a fish tank are about the same size, as the larger individuals will attack and eat any smaller tadpoles. If you intend to observe the development of more than one species, they should be kept separately.

Remember that newt tadpoles are more delicate than adult newts, if handled roughly, injured, or removed from water, even for a short time, they may die. Any newt tadpoles in the tank that appear distressed, suffer injury, don't feed or fail to develop properly should be returned to the pond as soon as possible.

56: July, three great crested newt tadpoles all about ten weeks old.
Drawn to small pieces of freshly chopped earthworm by the taste of blood in the water.

Newt tadpoles should always be released back into the pond before their external gills have been absorbed, as young newts they must have easy access to terrestrial habitats otherwise there is a possibility that they may drown.

All captive newt tadpoles, whatever their stage of development, should be returned to the pond by the end of August. It is not a good idea to try and keep any tadpoles or young newts in captivity through the winter because of the time and effort required to look after them properly in their terrestrial and over wintering phases.

Although a great deal is already known about our native newts, in my experience, there is always something new and interesting to discover about them. This is especially true now that newts have started to make their home in our ponds and gardens.

As more people take an interest in 'their newts' I believe that previously unseen or unrecorded behaviour will gradually come to light. By sharing your observations, experiences and records with others you can play an important part in advancing our knowledge of these fascinating creatures.

Chapter nine

What do you know about newts?

Newt quiz

The purpose of this newt quiz is for you to have a bit of fun answering the questions and to find out how much you know about our native newts. Following on from the quiz are some ideas and suggestions on how you can become more involved in newt conservation in your local area.

All the questions in the quiz are based on information to be found in this book, and it should take you about ten minutes to complete. To make marking easy and allow others to take the quiz use a blank sheet of paper for your answers. The correct answers can be found on page 87.

1 What is the best location for a newt pond ?

a sunny site **b** shady site **c** under trees

2 What sometimes happens to water in a new pond during the first couple of weeks ?

a it turns yellow **b** it turns green **c** it turns red

3 What is the area immediately adjacent to the pond known as ?

a the garden **b** the edge **c** the emergence zone

4 Duckweed on the ponds surface should be controlled by ?

a herbicide **b** a fine meshed hand net **c** twisting a cane in the water

5 Should fish and newts share the same pond ?

a yes **b** no

6 A place where terrestrial phase newts congregate during the day from April to October is called ?

a summer shelter **b** hidey hole **c** winter shelter

7 Good winter shelter is provided by ?

a flower borders **b** dry stone walling **c** overgrown lawns

8 Our largest native newt is ?

a smooth newt **b** palmate newt **c** great crested newt

9 Female newts of all species are always what ? than same age males.

a smaller **b** bigger. **c** thinner

10 Will newts of different species share the same pond ?

a yes **b** no

11 Which species of newt has poison glands in its skin ?

a palmate newt **b** great crested newt **c** smooth newt

12 Newt eggs are laid ?

a in clumps **b** in a string **c** individually

13 After laying newt eggs are ?

a covered in mud **b** wrapped in a leaf **c** left to float free

14 Egg laying reaches a peak in what month ?

a March **b** July **c** May

15 Newt eggs have a ?

a jelly shell **b** hard shell **c** no shell

16 Smooth and palmate newt eggs have a ?

a black nucleus **b** grey/brown nucleus **c** white nucleus

17 Great crested newt eggs have a ?

a grey nucleus **b** brown nucleus **c** white nucleus

18 In all newt species only the ? has a breeding crest.

a male **b** female **c** tadpole

19 The male palmate has ?

a an impressive breeding crest **b** no body spots **c** permanently webbed rear feet

20 In water the male smooth newt has ?

a webbed rear feet **b** feathered/splayed toes on the back feet **c** webbed front feet

21 The male great crested newt has a ? on his tail.

a white stripe **b** orange stripes **c** no stripe

22 The rate of newt egg development is influenced by ?

a day length **b** rainfall **c** temperature

23 Newt tadpoles take oxygen directly from the water using ?

a internal gills **b** external gills **c** lungs

24 Newt tadpoles emerge from the egg with fully formed and functioning ?

a legs **b** lungs **c** eyes

25 Newt tadpoles feed on ?

a algae **b** pond weed **c** live prey

26 In the pond adult newts feed on ?

a pond weed **b** live prey **c** algae

27 Will adult newts eat their own young ?

a yes **b** no

28 Great crested newt tadpoles develop poison glands at ?

a three weeks old **b** six weeks old **c** twelve weeks old

29 The majority of newt tadpoles start to absorb their external gills at ?

a six weeks old **b** twelve weeks old **c** twenty weeks old

30 Our smallest species of newt is ?

a the great crested **b** the smooth **c** the palmate

31 The majority of the season's young newts leave the pond in ?

a April/May **b** August/September **c** October/November

32 Newts move into winter shelters in ?

a August/September **b** December/January **c** October/ November

31 On land newts feed on?

a plants **b** insects and earthworms **c** detritus and carrion

34 Do some newt tadpoles, (of all three species), spend the entire winter in the pond ?

a yes **b** no

35 The majority of newt tadpoles leave the pond at ?

a four weeks old **b** sixteen weeks old **c** twenty weeks old

36 Newt tadpoles develop their ? legs first

a front **b** back **c** both sets

37 Can newt tadpoles re-grow a lost limb ?

a no **b** yes

38 What is the best way to catch very young newt tadpoles ?

a net **b** by hand **c** "dipping bowl"

39 The skin of the smooth and palmate newt on land is ?

a dry and leathery **b** smooth and slippery **c** rough and warty

40 The skin of the great crested newt on land is ?

a dry and leathery **b** smooth and slippery **c** rough and warty

41 Great crested newts are ?

a not protected by law **b** strictly protected by law **c** partially protected by law

42 Adult newts spend the majority of their life ?

a on land **b** in the pond

Answers

1. a 2. b 3. c 4. b 5. b 6. a 7. b 8. c 9. b 10. a
11. b 12. c 13. b 14. c 15. a 16. b 17. c 18. a
19. c 20. b 21. a 22. c 23. b 24. c 25. c 26. b
27. a 28. c 29. b 30. c 31. b 32. c 33. b 34. b
35. b 36. a 37. b 38. c 39. a 40. c 41. b 42. a

How to get involved in newt conservation

If you did well in the quiz and are interested enough, you might want to consider contacting your local Wildlife Trust, or one of the other wildlife organisations involved in amphibian conservation. Wildlife groups are always keen to recruit new members or enthusiastic volunteers and most would welcome any information or records you may have about previously unrecorded amphibian breeding sites in your area.

Organisations such as the Amphibian and Reptile Groups of the United Kingdom (ARG UK) and Froglife should be able to assist you with pro-forma amphibian recording and advice sheets, historical pond records, amphibian training events and contact with like minded people in your area.

Depending on your enthusiasm and experience these organisations might be able to help you obtain a survey licence from the relevant agency, enabling you to undertake great crested newt surveys, and perhaps get involved in local and national projects to help protect the species.

A good tip, from my time as a volunteer with the Pond Life Project, is to contact your local paper in spring with details about your role and the problems facing amphibians in your area. Most local papers are usually more than happy to print a topical spring time story about frogs, toads and newts. If the article includes your contact details you can expect to receive dozens of phone calls, and e-mails from members of the public with information about frog, toad and newt colonies in your area which you can follow up or pass on to your local wildlife group.

How involved you become and how far you decide to take amphibian conservation is entirely up to you. I started forty years ago catching newts at a local pond with a sixpenny net from the corner shop, and more by luck than judgement, and surprising no one more than myself, I ended up writing this book and secured a full time job working on amphibian conservation projects. So who knows what you could achieve with enthusiasm and a little luck!

I am always pleased to hear from others who share my interest in amphibians and happy to personally answer questions and give advice. To help facilitate this I have been working with others on a new website www.newtsinyourpond.com contact details and how to reach me are on page 92.

At the moment I am particularly interested in any reports of garden ponds supporting breeding populations of great crested newts (in strict confidence if necessary). I would also be interested in any information you might have on great crested newt eggs found in ponds between November and December, (this appears to be new or previously unrecorded behaviour, perhaps linked to climate change, and first reported by Julian Whitehurst in 2005). I would welcome via the website any information or comments you have on either subject.

I hope you have enjoyed my literary and photographic efforts and found this book interesting and informative. If you now create your own newt pond or get involved in amphibian conservation I will look on the time and effort taken to produce it as well spent.

Regards

Jim

Useful contacts

Natural England (NE)

Responsible for issuing great crested newt licences (England) that allow you to work legally with the species. They are interested in any information you might have on great crested newt colonies in your area. They produce a wide range of guidelines, handbooks, fact sheets, and leaflets relating to conservation issues and UK protected species, many are available as free downloads from their comprehensive website

Email: enquiries@naturalengland.org.uk

Website: **www.naturalengland.org.uk**

Tel: 01733 455000

Countryside Council for Wales (CCW)

The Countryside Council for Wales takes the lead in providing the public and business with information and advice relating to countryside and wildlife issues in Wales. They are the agency responsible for the issue of protected species licences in Wales. The CCW website is comprehensive, detailed and informative.

Email: Enquiries @ccw.gov.uk

Website: **www.ccw.gov.uk**

Tel: 08451 306229

Scottish National Heritage (SNH)

Scottish National Heritage is the agency responsible for countryside and wildlife issues in Scotland. As part of their work they issue all the protected species licences for Scotland. The SNH website is interesting and informative.

Email: enquiries@www.snh.org.uk

Website: **www.snh.org.uk**

Tel: 01463 725000

Amphibian and Reptile Groups of the United Kingdom (ARGs)

This organisation represents a growing number of local groups involved in a wide range of local and national reptile and amphibian conservation projects.

The ARG website will provide links to your local group and details of current projects. They are always pleased to hear from willing volunteers.

Website: **www.arg-uk.org.uk**

Wildlife Trusts

A useful source of information and involved with a wide range of conservation issues and projects. Many of them run amphibian training days and organise voluntary working parties to work on wildlife reserves and restore local ponds. They are always glad to hear from potential new members and willing volunteers, to contact your local trust search for wildlife trusts on the internet. Links to the other trusts can be found on the Cheshire Wildlife Trusts website.

Email: enquiries@cheshirewt.cis.co.uk

Website: **www.wildlifetrust.org.uk/cheshire**

Tel: 01948 820 728

Froglife

This organisation is involved in a wide range of amphibian conservation projects across the UK. They would be especially interested in any reports you have about unusual large-scale frog mortalities in your area. The froglife website is comprehensive, and their advice sheets and publications are interesting and informative.

Website: **www.froglife.org.uk**

Tel: 01733 558844

The British Herpetological Society

The British Herpetological Society is a learned society which undertakes conservation activities to benefit amphibians and reptiles, particularly British indigenous species. The Society is actively engaged in field studies and conservation management, and provides a platform for the open discussion of herpetology for scientists, conservationists and enthusiasts alike.

Website: **www.thebhs,org**

The Herpetological Conservation Trust

The HCT is national charity that manages 80 nature reserves around the country. The trust focuses on the conservation of reptiles and amphibians through practical conservation management. They run training events and give advice on all aspects of reptile and amphibian conservation.

E-mail: enquiries@herpconstrust.org.uk

Website: **www.herpconstrust.org.uk**

Tel: 01202 391319

Aqualife

Owned and run by David Hatter a passionate and knowledgeable advocate for our native pond and wetland plants. Newly opened to members of the public in 2006 this commercial nursery produces a wide variety of native pond and wetland plant species, available in large or small quantities wholesale, retail or grown to order for major conservation projects. The nursery provides a range of design construction and maintenance services, Domestic, Industrial, Commercial and Public. Always interesting and helpful David will happily give advice to phone and email callers and a warm welcome to visitors at the nursery.

Email: info@aqualifeltd.co.uk

Website: **www.aqualifeltd.co.uk**

Tel: 01625 875333

Landlife

Are a registered environment charity, working mainly in urban and urban fringe areas, to bring nature and people closer together. They grow and supply native wildflower seeds and plants in small and large quantities to anywhere in the UK mainland. They can help you to choose the right species for your site, give advice on how to prepare the soil and tips on managing your wildflower area successfully.

E-Mail: info@landlife.org.co

Website: **landlife.org.co**

Tel: 0151 737 1819

newtsinyourpond.com

This website has been set up to help promote newt conservation across the UK. The aim is to raise public awareness of the problems newts face in the modern world and show how individuals can help ensure the survival of these fascinating creatures.

The website gives Schools, Wildlife Groups, Businesses and Individuals access to a wide range of features and services including:

- **School Visits:** Arrange a visit to your School from our licensed newt expert with interactive amphibian exhibit.

- **Organised Events:** Amphibian exhibit, for your conservation event, with a licensed newt expert on hand throughout the day to answer questions and give advice.

- **Speakers:** Book a speaker for your group or society for entertaining talks, supported with live specimens, about frogs, toads, newts or wildlife gardening.

- **Pond Surveys:** Have a licensed newt expert visit your site, to survey for great crested newts, and advise you on the legal implications of having them on your land. Confidential reports prepared, commercial and domestic work undertaken.

- **Questions and Answers:** Your amphibian questions answered online by our experts.

Email: enquiries@newtsinyourpond.com

Website: www.newtsinyourpond.com

Tel: 01625 869921